ENERGY AND CLIMATE CHANGE

ENERGY AND CLIMATE CHANGE

An Introduction to Geological Controls, Interventions and Mitigations

MICHAEL STEPHENSON

Elsevier
Radarweg 29, PO Box 211, 1000 AE Amsterdam, Netherlands
The Boulevard, Langford Lane, Kidlington, Oxford OX5 1GB, United Kingdom
50 Hampshire Street, 5th Floor, Cambridge, MA 02139, United States

Library of Congress Cataloging-in-Publication Data
A catalog record for this book is available from the Library of Congress

British Library Cataloguing-in-Publication Data
A catalogue record for this book is available from the British Library

ISBN: 978-0-12-812021-7

Publisher: Candice Janco
Acquisition Editor: Amy Shapiro
Editorial Project Manager: Tasha Frank
Production Project Manager: Prem Kumar Kaliamoorthi
Cover Designer: Victoria Pearson

Typeset by SPi Global, India

DEDICATION

For Fred and Jack.

QUOTATION

If we want things to stay as they are, things will have to change.
(Se vogliamo che tutto rimanga come è, bisogna che tutto cambi)
From Il Gattopardo (The Leopard), 1958
Giuseppe Tomasi, Duke of Palma di Montechiaro and Prince of Lampedusa

CONTENTS

PREFACE

Climate change and energy are rightly connected in the sense that many of the ways that humankind generates energy on an industrial scale involve burning fossil fuels. This is well known and in the early part of this century we are at last trying to do something about this link that leads to global warming. What is less well known is that the larger geological part of the carbon cycle, the cycle that traces carbon's route through the atmosphere, biosphere and geosphere, is also capable of altering climate in the opposite way: lowering temperature by removing carbon dioxide from the atmosphere.

Climate change and energy work within the boundaries of Earth system science and can therefore be modelled mathematically, allowing aspects of the larger changes in the environment to be forecast. Large changes can also be seen in the long history of the Earth, some large enough to cause mass extinctions and perturbations in the history of life, such as mass die-offs that happened about 250 million and 65 million years ago. These perturbations are enough to make us sit back and think about the Earth's fragility and the fact that, for the first time in the history of the Earth, humankind is able to alter the delicate balance of the planetary system. To mark this, a new geological epoch known as the Anthropocene is being proposed by geologists.

But our understanding of the slow turning of the geological part of the carbon cycle does not allow prediction at timescales that allow policy makers and planners to prepare for the local and imminent changes that climate change will make, nor provide answers to how we adapt to climate change or mitigate it. For this we mainly turn to extrapolation of the surface processes in the weather, ocean and ice caps that allow predictions and plans to be made. Understanding these processes allows us also to see where non-linear changes may take place, for example at tipping points, or when positive feedback loops accelerate change that becomes runaway change. These feedback loops and tipping points need not all be natural scientific processes, but can relate to human opportunism and serendipity, for example the historical proximity of early oil production and automobile manufacture, or the take up of coal to power steam pumps so that coal mining could become more efficient.

As well as layers of rocks providing the story of the long-term carbon cycle, they also act as physical materials to provide some of the ways that humankind can survive climate change. In a short circuit of the natural sequestration part of the geological carbon cycle, CCS or carbon capture and storage could allow us to remove industrially generated CO_2 from fossil-fuel power stations, refineries, cement factories and steel works. Bioenergy and CCS (or BECCS) could provide a method of achieving net negative emissions. Geothermal power emits no CO_2, and natural gas and compressed air energy storage might help solve the intermittency problem caused by renewables such as wind and solar.

For many—and perhaps particularly the poorer people of the world—groundwater may help to buffer the effects of climate change on surface water in rivers and lakes, helping to future-proof against water scarcity for agriculture, industry and growing cities. The underground may also be the place in cities where excess floodwater can be stored rapidly out of harm's way.

Perhaps the most interesting and fundamental change on the horizon is the fact that although humankind can change the Earth for the worse, it can also reverse those changes, first by understanding the processes of change and then by doing something deliberate and conscious to alter that change. We are already making small changes. Meteorologists noticed the depletion of atmospheric ozone, and humankind reacted by making policies that stopped further depletion. Non-smokeless coal was phased out in London in the 1950s to stop dangerous smog; coal was phased out in Pittsburgh steel works in the 1880s to improve the city's atmosphere; and sulphur began to be captured from coal power station flue gas in the 1970s to reduce acid rain.

This intrusion of human conscious effort is a new teleological kind of feedback in the climate and energy system. For it to work properly requires a deep understanding of physical processes and how they work at human timescales: days, minutes, and hours. This means that natural processes need to be measured. The surface of the Earth is being measured in unprecedented detail using a range of sensors and new telemetry so that data can be compiled, analysed and displayed on the fly. Meteorologists use data of this type to develop weather forecasts which allow us to properly manage agriculture, transport and other activities; air quality is also measured. Increasingly the ecology of natural environments is being measured: river discharge, soil temperature and moisture content, both by satellite and by in situ sensors. The distribution, fertilisation and yield of crops can now be monitored by satellite and altered by planting machines and harvesters guided by high-resolution GPS.

The part of the natural world that is not measured enough is the subsurface. It is true that volcanism and earthquakes are tracked at high geographical resolution and in real time, mainly to manage the hazard and related risk to human life and property; but the kinds of 'state-of-the-environment' measuring and monitoring that are commonplace above the surface are not carried on below. In some parts of the world groundwater levels are measured, but there is not much monitoring of other properties of groundwater such as temperature, salinity, pH and gas content. This is important in areas where perturbations of the subsurface are likely—where gas is stored, large-scale manufacturing is done, or where oil and gas are extracted—but particularly where there are multiple and competing uses of the subsurface. These perturbations will only increase as the developing world grows. In the same way that we try to manage our activities with respect to the weather, we need to measure, monitor and manage the subsurface. This is geological science, but not the geological science that many geologists were trained in. Rather than the age of rocks and the story that rocks tell about Earth history, this is the rock mass as a container and a conduit through which important fluids flow—water, methane, carbon dioxide. The geologists that consider these rock masses as a stage on which environmental processes play out are more likely to see rock, and soil above it, as a very variable material with multiple properties that change in time, perhaps over minutes and hours—and that change geographically and with depth over metre scales. These new geologists might be able to forecast landslides and groundwater flooding in the same way that a meteorologist forecasts weather. They might be able to detect, limit and control subsurface leaks of fluids or dangerous pressure changes. They would likely know the physical properties of Cretaceous and Jurassic rocks, rather than their ancient origin.

The technology to enable this change in geological science is becoming available: sensors designed for the tough subsurface environment, 3D visualisation software, computing power and telemetry. With the take up of more technology, geological science will become more quantitative and less descriptive, and therefore better able to deal with risk and hazard, liability and law.

ACKNOWLEDGEMENTS

Thanks to Ruth O'Dell, Joel Gill, Mike Ellis, and Dave Schofield for commenting on the text, and to the staff of the British Geological Survey and to other scientists that I have met in the last few years while planning and writing this book. Even with all this expert discussion and advice, however, any errors are entirely my own.

NOTE TO THE READER

This book is meant as an introduction to a number of connected issues in energy and climate change, including the carbon cycle, climate change in deep time, the 'fossil economy', the role of geology in climate change abatement and adaptation, and the importance of geological measurement and system understanding. Chapters aim to establish these connections, rather than delve deeply into detail. For further detail the reader is referred to the bibliography of sources given at the end of each chapter.

The word *energy* as used in the book is broader than the physicist's definition of the natural phenomenon. Here, energy has a broad meaning including the chemical energy in fossil fuels, but also the industry and processes relating to energy supply. Necessarily this means reference to many different kinds of units of measurement, and these are mixed in the book, reflecting different sources of information, for example scientific papers, reports, and books. To help in simple comparison between units of measurement, I have included a conversion table at the back.

CHAPTER 1

The Carbon Cycle, Fossil Fuels and Climate Change

Contents

Geological science offers a unique way of looking at the relationship between energy and climate change. The neatest way to see this connection is through the carbon cycle—the path that carbon takes through the atmosphere, biosphere and geosphere—and to see the limits to that cycle, and the controls on its rate and character that are subject to natural fundamental laws. These laws can be modelled through the new discipline of Earth system science, revealing the extent to which life in all its forms interacts with other big forces to change the world. This is best illustrated in how the carbon cycle is at the centre of the tension between our use of energy and the atmosphere, and the geological generation of much of the fossil fuels we use to generate power. Carbon leaves the atmosphere largely through the engine of life, including biological pumps operating in the ocean. In a permanent form below the surface of the Earth, essentially as fossil fuels like coal, gas and oil, this carbon can do no harm. Of course, carbon can also leave the rocks and get back into the atmosphere. For example, carbon dioxide is released during the metamorphism of carbonate rocks when they are subducted into the Earth's mantle, or by volcanoes. But it is humankind's intervention in the carbon cycle, a short cut within this large geological part of the carbon cycle, which is causing the problem: the burning of fossil fuels.

Energy and Climate Change
https://doi.org/10.1016/B978-0-12-812021-7.00001-4

A lot of us are now familiar with the famous photograph taken of the Earth on February 14, 1990 by the Voyager 1 space probe from a distance of 6 billion kilometres. In the photograph, the Earth appears as a pale blue dot against the blackness of space. During a lecture at Cornell University in 1994, the cosmologist Carl Sagan showed the image to the audience and contemplated the deeper meaning of the pale blue dot, in relation to the ultimate negligibility of humankind against the vastness of space. But the pale blue dot also helps us to realise the physical, chemical and biological boundaries that limit our planet. Most of the processes that I talk about in this book have limits, rates and thresholds that are governed by a system. The science of this is called Earth system science. It is fairly new and it is very interdisciplinary, using elements of geology, physics, chemistry, biology and mathematics.

One of the most interesting early findings of Earth system science is the extent to which life in all its forms interacts with other big forces to change the world. In a sense, the tension between our use of energy and the atmosphere and the geological generation of much of the fossil fuels we use is one of the best demonstrations of the way in which life interacts with other big forces, as I hope you will see. This is a unique geological way of looking at the problem of energy and climate change.

THE CARBON CYCLE

Geologists were amongst the first to recognise that life has had a powerful role in shaping the Earth. James Hutton, one of the founders of geology, described the Earth as '…not just a machine but also an organised body, as it has a regenerative power…' The Russian geologist Vladimir Ivanovich Vernadsky was one of the first geologists to hypothesise that life is a geological force that shapes the Earth, suggesting that the oxygen, nitrogen and carbon dioxide in the Earth's atmosphere result from biological processes. During the 1920s he published works arguing that living organisms could reshape the planets as surely as any physical force.

At the heart of climate change and energy is the carbon cycle, which involves exchange of carbon between 'stores' or accumulations in the atmosphere, terrestrial biosphere, oceans, and the subsurface of the Earth (Fig. 1.1). The rates of exchange and the sizes of the stores are instrumental in the ability of the Earth to sustain life. The carbon cycle also interacts with

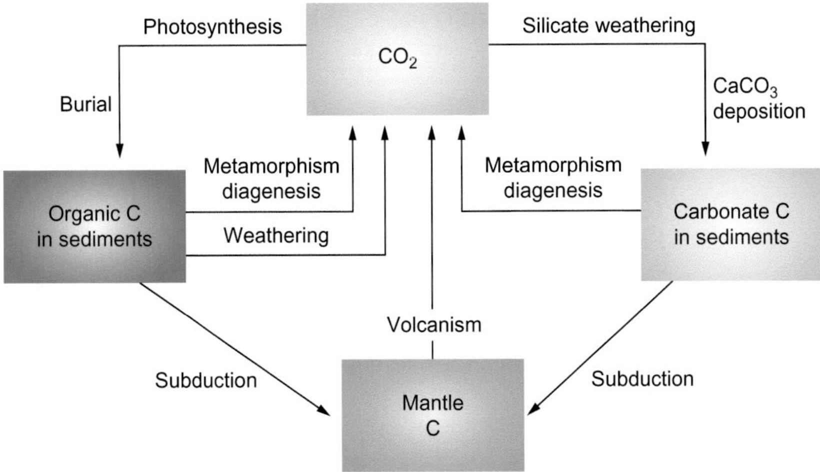

Fig. 1.1 The carbon cycle. On the left side photosynthesis creates organic matter that collects in sediments (carbon burial or capture). This carbon can remain out of the atmosphere for a very long period locked away in rocks for millions of years. But eventually the carbon can be returned through weathering of exposed rock. It can also—through subduction at plate tectonic boundaries—become part of the mantle. The mantle might return it through volcanoes. *(From Berner, R., 2003. The long-term carbon cycle, fossil fuels and atmospheric composition. Nature 426, 323–326.)*

other very large cycles of oxygen, water, phosphorous and nitrogen in complicated ways that are difficult to model and predict.

In a book purely about climate change, the main parts of the carbon cycle story would be concerned with the first three of the stores: the atmosphere, terrestrial biosphere, and oceans, but in this book the addition of energy means we need to look at one in detail: the deep Earth, which supplies human society with most of its energy and fuel.

Broadly this last part of the carbon cycle is geological rather than atmospheric or biological. It is distinct in that it operates over very long periods—millions or tens of millions of years. These time periods may seem academic to human society, but in fact are the periods that are needed to accumulate carbon such as coal, gas or oil—and in the long run also affect the amount of carbon dioxide in the atmosphere and therefore climate.

Though the immediate non-geological exchange of carbon is most important to climate change on human timescales, most of the Earth's carbon is actually stored below the surface (Table 1.1); and a lot of this resulted from living things at the surface, either directly or indirectly. Eighty percent of this is limestone (sedimentary carbonate), from sedimentation

Table 1.1 Stores of Earth carbon. The big stores are in the lithosphere (the rocks)—sedimentary carbonates and kerogen

Pool	**Quantity (gigatons)**
Atmosphere	720
Oceans (total)	38,400
Total inorganic	*37,400*
Total organic	1000
Surface layer	670
Deep layer	36,730
Lithosphere	
Sedimentary carbonates	>60,000,000
Kerogens	15,000,000
Terrestrial biosphere (total)	2000
Living biomass	600–1000
Dead biomass	1200
Aquatic biosphere	1–2
Fossil fuels (total)	4130
Coal	3510
Oil	230
Gas	140
Other (peat)	250

Source: Wikipedia retrieved July 2017.

from seawater of calcium carbonate or shells, and the rest is organic matter from buried dead organisms. We will look at the way that carbon makes its way from the 'living world' into the rock in the next section.

But before that we need to investigate the surface parts of the carbon cycle. The exchange of carbon between the atmosphere and the land is largely a biological process with plants photosynthesising to make carbon compounds, and aerobic respiration releasing it again—though fires can also do this. The exchange between the atmosphere and the sea involves chemical processes such as dissolution and marine photosynthesis working as solubility and biological 'pumps' that take carbon from the atmosphere mostly to the deep ocean. Marine animals that produce shells also extract carbon from seawater to allow calcium carbonate to lock up carbon in the long term. Also long term is the process of silicate weathering on land where rocks containing silicates take up carbon dioxide to produce calcium carbonate and SiO_2.

Carbon can leave the rocks in several ways. Carbon dioxide is released during the metamorphosis of carbonate rocks when they are subducted into the Earth's mantle. This carbon dioxide can be released into the atmosphere and ocean through volcanoes. It can also be removed from rocks by humans through the burning of fossil fuels.

HOW FOSSIL FUELS ARE FORMED

As I have already indicated, fossil fuels are forms of organic carbon in sedimentary rocks, mainly coal, oil and natural gas.

It is hard to overemphasise the importance of fossil fuels in modern energy. A quick eyeball of Fig. 1.2 shows the dominance of fossil fuels (coal, oil, natural gas) in primary energy consumption in the world, mainly in transport, heating homes and for generating electricity. Fig. 1.3 shows the importance of fossil fuels in electricity generation alone worldwide. The right-hand side shows that, though the amount of electricity generated in the world has gone up hugely from 6 TWh (terawatt hours) to nearly 24 TWh between 1973 and 2014, the way that it is generated has not changed much. Put simply, coal is still the backbone of global electricity.

The formation of fossil fuels is covered in the left-hand side of the carbon cycle shown in Fig. 1.1 and it involves an exchange from the atmosphere to the land (in coal) or to lakes and the sea (in oil and natural gas, or methane hydrates). Though this exchange is continual (in other words fossils fuels are

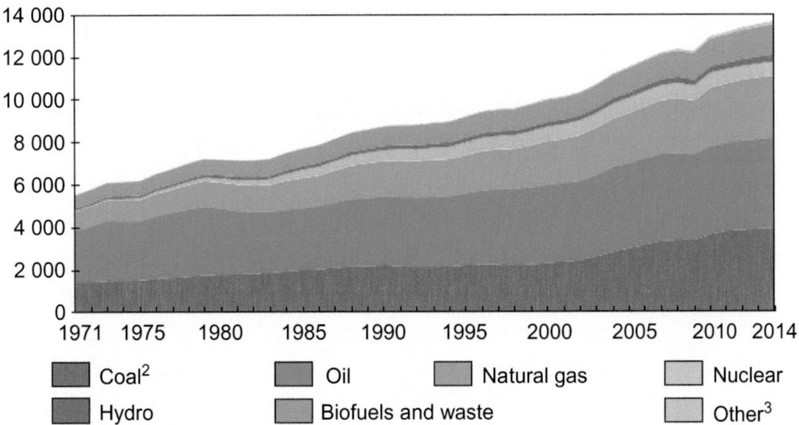

Fig. 1.2 World total primary energy supply from 1971 to 2014 by fuel (millions of tonnes of oil equivalent, Mtoe). *(From Key World Energy Statistics, International Energy Agency 2016. Key World Energy Statistics. https://www.iea.org/publications/freepublications/publication/KeyWorld2016.pdf.)*

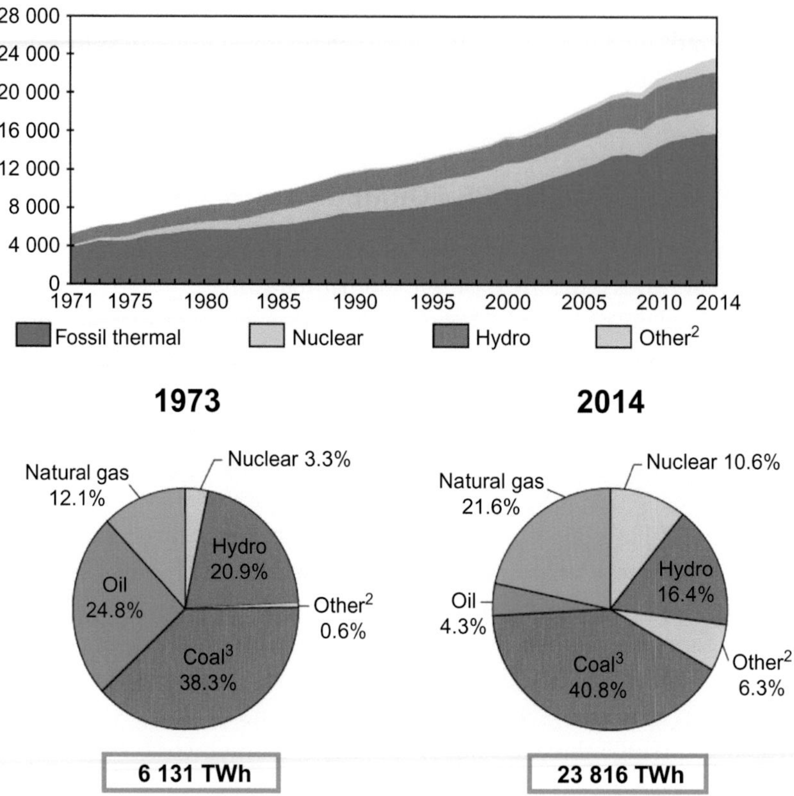

Fig. 1.3 World total electricity production by fuel (terrawatt hours TWh). *(From Key World Energy Statistics, International Energy Agency 2016. Key World Energy Statistics. https://www.iea.org/publications/freepublications/publication/KeyWorld2016.pdf.)*

continually being made), their use by modern society is so many more times faster than they are replenished that they are effectively non-renewable.

Coal

Coal is a sedimentary rock usually occurring as thin layers called coal beds or coal seams. Coal is formed on land, but under conditions of high water table in standing water in swamps and marshes. Much of Europe's and North America's coal was formed around the same time during the Carboniferous period (360–300 million years ago), though other periods such as the Permian, Jurassic and Cretaceous also witnessed coal formation.

The carbon in Carboniferous coal was mostly produced in lowland forest dominated by plant types including ancient relatives of the club mosses, horsetails, ferns, seed ferns and conifer-like trees. Essentially, the process of photosynthesis locked carbon (from CO_2 in the Carboniferous atmosphere) into plant tissues, the plants died and the remnants of tissues were preserved in soil and sediment—and then rock. This is the short version, but details of how this happens is a story of several factors conspiring to store or capture carbon. Without all these, carbon does not get sequestered.

It is interesting to note that the swamp forests of the Carboniferous were a relatively new feature of the land 360 million years ago. Although plants were well established on some parts of the land, large areas were probably still uncolonised. Fossil evidence for the earliest land plants is limited to spores from moss or hornwort-like plants 475 million years old, 175 million years before the coal swamp forests. The whole plants, when they eventually do appear as fossils, show a simple vascular system that had tissues for conducting water, minerals, and photosynthetic products through the plant. This also meant that they had a certain rigidity (from vascular pressure) and so could reach up for sunlight, though not very far. The earliest known of these plants, *Cooksonia* (mostly from the northern hemisphere) and *Baragwanathia* (from Australia), were only very small, millimetres high at the most. By 400 million years ago, primitive plants had created the first recognisable soils that harboured mites and scorpions. Strangely these plants did not have leaves; small leafless shrubs filled the landscape until the tree-like fern *Archaeopteris* appeared with the first leaves. By the start of the Carboniferous period plants and trees with leaves were common and the first seed-forming plants had appeared. It also appears that a number of plant types were able to produce lignin, whose main function was to strengthen wood (or xylem cells) in the stems of trees so that they could grow taller—at least taller than their competitors. So trees could grow taller and become more efficient photosynthesisers. Lignin also has relevance for the fossil fuel story because it is more persistent than other plant tissues and preserves better.

Extinct, huge forms of lycopsids (club mosses) dominated the Carboniferous coal swamps. Their remains are well known as the fossil *Lepidodendron* with its distinctive diamond-shaped 'scales' (Fig. 1.4).

The diamond-shaped repeated patterns are 'leaf cushions'—and these, like the leaves, were capable of photosynthesis because both preserve traces of stomata (the pores which allow CO_2 to diffuse into plants for photosynthesis). This ability to photosynthesise over large areas meant that lycopsid trees like *Lepidodendron* grew fast to heights of 50 m with a trunk diameter of over 1 m and a rooting system spread over an area with a diameter of 24 m.

Studies of the sediments surrounding the fossils reveal that *Lepidodendron* peat-forming forests lived alongside horsetail and seed-fern riverbank forest,

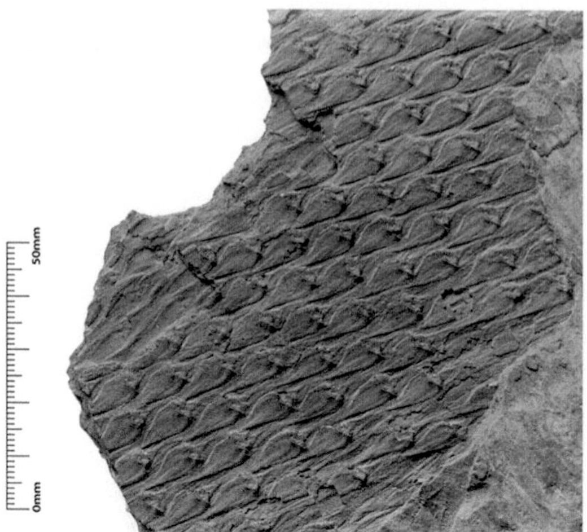

Fig. 1.4 An example of *Lepidodendron* from the collection of the British Geological Survey. (Lepidodendron glincanum *Eichwald, River Esk, Canonbie, BGS image P686121. BGS copyright NERC.)*

all growing in semiflooded conditions, with animals including molluscs, arthropods and tetrapods—including the earliest-known reptiles. Inland were drier plains of cordaite forests (a primitive conifer-like plant) that were prone to wildfire. Coal swamps persisted for single periods of hundreds to thousands of years, where peat accumulated at the same rate as water table rise, so that the best conditions for preservation were created. Each of these periods effectively created a single coal seam.

The coal swamps stretched across what is now North America, Europe and Asia at the height of their development about 300 million years ago and led to the coal deposits of the Northeast and Midwest of the United States, the Maritimes of the United States and Canada, Britain, northern and eastern Europe and Asia. These were the fuels that much of the industrial revolution was built on.

From Coal Swamps to Coal

But for coal to form—and for carbon to successfully transition from the biosphere to the geosphere—the carbon of those swamps had to find its way into rocks. This began with the partial preservation of the coal swamp carbon as peat. Peat preserves best, in the long term, in places with low elevation that are partially flooded and are slowly subsiding.

Today peat covers 3% of the Earth's surface and grows in many environments where accumulation of dead plant material exceeds the rate at which it is removed by atmospheric oxidation, burning or erosion. Modern peat-forming environments are known as 'mires' and a 'bog' is a mire which mainly gets its water from direct rainfall, whereas a 'swamp' is a mire at relatively low elevation which receives its water from the underlying water table. Bogs often form at high elevation in hills and mountain areas but have a rather low long-term chance of preservation. Swamps have better preservation potential in the sense that their carbon is more likely to find its way into rocks through the long-term carbon cycle.

Swamps often start by infilling a lake or other poorly drained lowland hollow. Mud, sand and plant debris slowly fills the hollow up. In climates where rainfall is high throughout the year, plants grow fast. The water table remains high and stagnation allows organic material to rot only very slowly, so the peat becomes thicker and may begin to form a very low dome because year after year more plant debris is added. The swamp therefore begins to rise above the surrounding land slightly. In this way the peat-forming process becomes more pure because it is more difficult for sediment from outside to contaminate the peat. The swamp may also become a bog because the primary source of its water may be rainwater if the water table does not reach high enough to slow the rotting of the vegetation. Such 'raised' swamps and bogs probably provide the highest quality coal because the coal does not contain sand or mud, only relatively pure carbon.

But even at this late stage in the transition from the biosphere to the geosphere, peat can still be dispersed. For example, catastrophic weather, like a hurricane, may be enough to erode a coastal raised swamp in a few days, in which case the plant debris that has collected would be scattered by the sea.

Long-term sequestration of carbon in peat happens easiest in areas where the land is subsiding. If the rate of subsidence is right, then peat can form for a long period and a final thick inundation with new sediments caps and seals the peat and the peat begins its burial process. More and more layers of other sediments cover the layer and 'coalification' begins.

Coalification

After burial, plant debris is geochemically altered by heat and pressure over a very long period of geological time. Heat is the most important variable, and if it is available in a basin (for example from nearby volcanic activity) relatively young coal can be generated. But in most cases the temperature increase that applies is the mean geothermal gradient of the Earth's crust, which is about 25°C per 1000 m of descent. Most bituminous coal forms at temperatures of 100–150°C. The quality or 'rank' of the coal increases with depth. In the bituminous coal stage the organic material is heated to a point where hydrogen-rich compounds generate jelly-like bitumen which

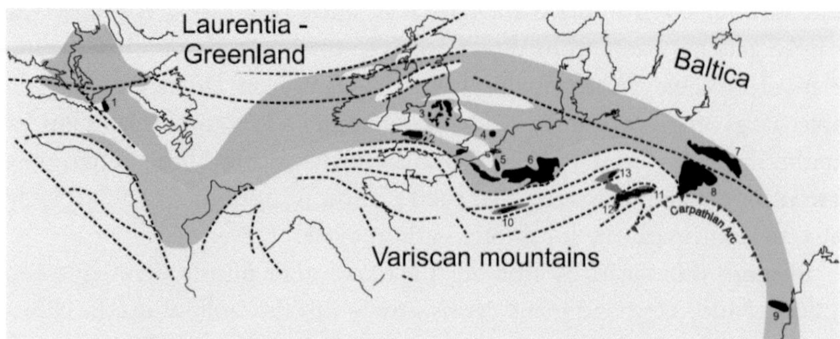

Fig. 1.5 A reconstruction of the Variscan Mountains at about 300 million years ago. The dashed lines show the positions and trends of the main folds and faults. Light green—foreland basin; grey-green—cratonic basin; black—major coalfields. *(From Cleal, C.J. et al., 2010. Late Moscovian terrestrial biotas and palaeoenvironments of Variscan Euramerica. Netherlands J. Geosci. 88, 181–278.)*

fills pore spaces in the coal. At this point the coal becomes denser and less porous. Further heating cracks the bitumen down to smaller molecules such as carbon dioxide and methane. Methane adsorbs onto the surface of organic matter or exists as a free gas in pores in the coal. This is known as coal bed methane.

As well as having coal swamps, the Carboniferous of Britain, North America, and eastern Europe also experienced very long periods of geological subsidence associated with a period of mountain building known as the Varsican Orogeny (Fig. 1.5).

Nowadays the remains of this mountain building can be seen in the hard metamorphic rock of the southwest of England and northern France, the Appalachian Mountains of the northeastern United States, and in the Hartz Mountains and Black Forest in Germany. But between 380 and 280 million years ago, the Varsican Orogeny created a range of very high mountains across northern France and Germany. At the near edge of such mountains, subsidence can be very intense, mainly due to the bulk and weight of the nearby mountain range sinking downward and pulling the Earth's crust with it. This area of strongly subsiding land in front of the mountains is known as a foreland basin, and this created the conditions for the coal swamps of the later continents of Europe and North America to be preserved for posterity.

Oil and Natural Gas

We have looked at the creation of coal. A similar and parallel process ensures that organic matter generated in the biosphere from aquatic organisms like algae and plankton finds its way into rocks. As with the formation of coal, several factors have to be in place or the organic matter will not become oil or gas. Again, these processes are part of the natural slow-moving geological carbon cycle where an oxygen-deficient environment allows the organic matter to be captured in the first stage of the long route underground. In this, it is no different from coal in that swamp water lacks the oxygen that allows organic matter to be consumed and ultimately returned to the atmosphere. In the case of oil and gas, though, the basic environment is different—the deeper water of seas, oceans and lakes. The different environment means that we start with a different biology—the micro- and macroscopic occupants of seas and lakes, not the plants of a terrestrial environment. If the organic matter gets preserved in lakebed or seabed sediments, then after burial under other sediments the heat and pressure begin to 'cook-up' the organic material so that some of it turns into gas or oil. We will look into the details of how this happens later, but for now we will concentrate on the fact that oil and gas have been created. If conditions are right, then oil and gas can get out of the rock in which they have been created; then the oil and gas tend to migrate upward because they are more buoyant than the fluid that usually saturates rock (water or brine). They find their way into sandstone and limestone, which soak them up like a sponge (called a 'reservoir' rock) and where they cannot rise any further because of a particular rock structure. This structure is called a trap and the trap allows accumulations of oil and gas big enough to make it worthwhile drilling from the surface to get them.

But understanding the actual process of preservation is important because it illustrates that, like coal, oil and gas do not get into the rocks unless many factors are satisfied. In many ways much carbon in rocks is an accident of history or geology. Paradoxically, the characteristics of the organic matter and the form of the rocks that enclose the organic matter both come together to allow geologists that study the climates of the past (palaeoclimatology) to reconstruct how climate has changed in the past. We will look at this in a later chapter.

Primary Production

The dominant biological sources of organic matter in the oceans and lakes are phytoplankton, algae, land plant debris and bacteria. Though the larger inhabitants of the seas do contribute organic matter when they perish, there is much more biomass in the microscopic organic matter. Antarctic krill (a tiny, swimming crustacean), for example, has an estimated global biomass of 379 million tonnes while the cyanobacteria are considered to have a global biomass of 1000 million tonnes. The distribution of this biomass within the oceans depends on the presence of nutrients such as nitrogen and phosphorous.

The map (Fig. 1.6) shows the gross distribution of modern photosynthesising plants in the ocean and on land, showing concentration related to latitude, ocean currents and terrestrial run-off (concentration of organisms close to river estuaries and deltas). Though the land and ocean arrangement would have been different at times in the geological past, latitude, ocean currents and terrestrial run-off would still have had an influence.

Preservation

Living things die and in the oceans this tends to mean that once-living tissue begins to fall through the water column to the seabed. The first link in the chain to oil and gas is thus for the tissue to survive the fall through the water column and then the entombment in the sediment at the sea bottom. At any stage the material could be consumed by another organism as food or be oxidised back to carbon dioxide and ultimately get back to the atmosphere.

What stops this happening is a rapid fall to the seabed, so that there is not much time for carbon to be oxidised, and a quick burial in ocean bottom sediment. It also helps if the ocean bottom sediment is fine grained. Mud is good because its small particle size impedes the movement of oxidants such as dissolved oxygen and sulphate, which slows the degradation process down.

Perhaps the most important factor that encourages the preservation of the fallen organic matter, though, is anoxia. From its name, it is clear that this means ‘in the absence of oxygen’, meaning dissolved oxygen in the water. Often some of our modern, more enclosed seas such as the Black Sea and Baltic Sea have anoxic water deep down or at the seabed. This is usually because oxygen cannot penetrate to those levels because the rest of the water above acts as a barrier. Deep anoxic parts of modern and ancient seas are (or were) good at ensuring that the remains of marine organisms are preserved in the sediment.

Maturation

What happens next is a slow process of biological, physical and chemical alteration of organic matter into oil and gas, and this is known as maturation. Rock that contains organic matter that has turned into oil and gas in the interstices between the rock particles is known as ‘mature’.

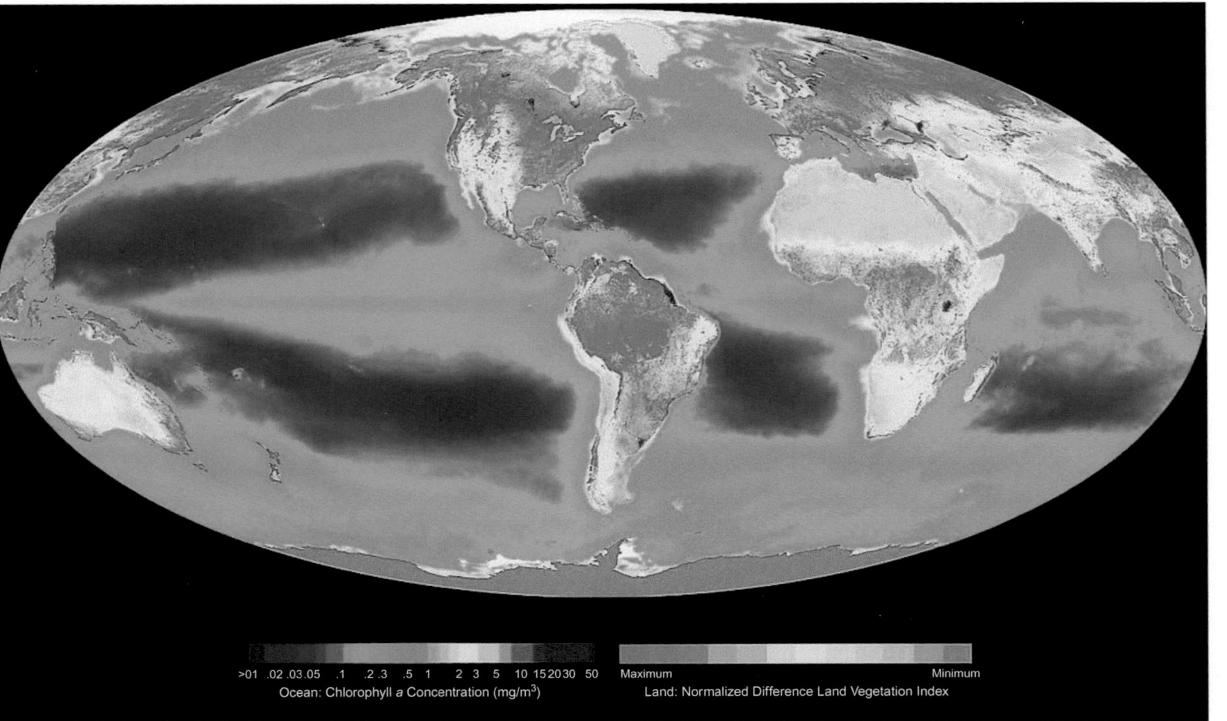

Fig. 1.6 The distribution and concentration of a large component of the modern ocean and terrestrial biomass, basically photosynthesising plants in the ocean and on land. *(From the SeaWiFS Project, NASA/Goddard Space Flight Center and ORBIMAGE.)*

Maturation begins as soon as the layer of organic-rich sediment is buried under the next layer. A series of low-temperature reactions that involve anaerobic bacteria reduce the oxygen, nitrogen and sulphur in the organic matter, leading to an increased concentration of hydrocarbon compounds. This continues until the sediment reaches about 50°C. After that, the effect of high temperature increases the reaction rates, and the solubility of some of the organic compounds increases.

Heating naturally happens with burial; the temperature reached at a given depth depends on the geothermal gradient, as discussed briefly in the previous section.

The simplified diagram (Fig. 1.7) shows the track of the burial of organic matter with mixed ocean and terrestrial origin through a typical geothermal gradient. Large amounts of oil and gas only begin to form at temperatures over 50°C and the largest quantity is formed at between 60°C and 150°C. At still higher temperature, oil becomes thermally unstable and breaks down or 'cracks' to natural gas.

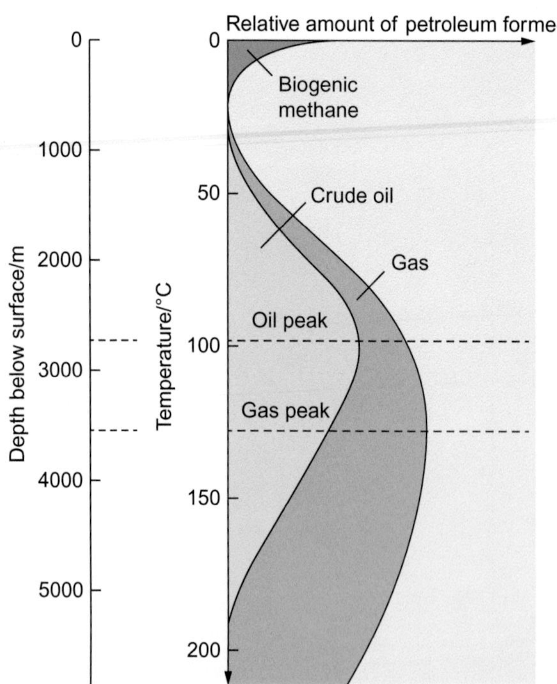

Fig. 1.7 The relationship between depth of burial, temperature and the relative amount of crude oil and natural gas formed from Type II kerogen in an area with a geothermal gradient of about 35°C/km. *(From the Open University website, http://www.open.edu/openlearn/ocw/mod/oucontent/view.php?printable=1&id=4763.)*

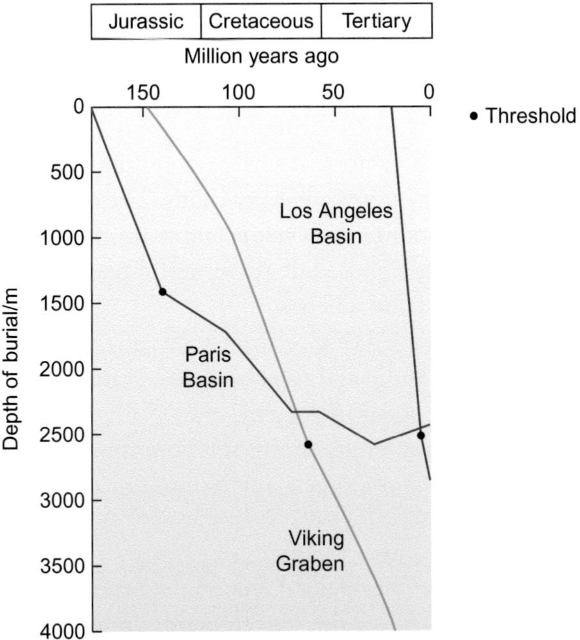

Fig. 1.8 Burial histories in three geological basins: the Los Angeles basin, the Paris basin and the Viking Graben in the North Sea. *(From the Open University: http://www.open.edu/openlearn/ocw/mod/oucontent/view.php?printable=1&id=4763.)*

The point at which oil and gas is first generated is known as the threshold. The length of time it takes to get to this point depends on how long it takes to heat up and this depends on how fast the layers are buried, like coal. This varies in different areas. The diagram (Fig. 1.8) shows how this happened in the Paris Basin, the North Sea Viking Graben and the Los Angeles Basin. This was reached after 40 million years in the Paris Basin (i.e., about 140 million years ago) when rocks 175 million years old were buried to a depth of 1400 m. It took 80 million years before the 150-million-year-old rocks in the Viking Graben started to generate oil and gas.

Chemistry and Physical Form

Some generalisations can be made about the chemistry of oil and gas. Essentially they are a complex mixture of hydrocarbons and lesser quantities of other organic molecules containing sulphur, oxygen, nitrogen and some metals.

In hydrocarbons, the number of carbon atoms in a compound determines its physical properties. Simple compounds such as methane (one carbon atom), ethane (two carbon atoms), propane (three) and butane (four) all have boiling temperatures below 0°C and are therefore gases at the surface.

Larger, more complex hydrocarbon compounds ranging from pentane to hexadecane are liquids at normal surface conditions. Even larger compounds with many carbon atoms are waxy solids. The commonly occurring compounds known as paraffins and cycloparaffins are chemically stable, with simple carbon-to-carbon bonds, but there are also more complex compounds such as aromatics and alkenes.

These compounds mix up to give three main physical forms for hydrocarbons: natural gas, crude oil, and solid bitumen. Natural gas is mainly gaseous hydrocarbons with small molecules; crude oil is a liquid with a very wide range of hydrocarbon compounds; solid bitumen contains the largest molecules that often contain oxygen, sulphur and nitrogen.

Migration

After oil and gas leave the fine-grained sedimentary rock in which they have been formed, buoyancy moves them from depth up towards the surface of the Earth, because it is less dense than the water or brine that occupies the spaces within the rock. They continue to migrate upwards until they are trapped beneath an impermeable rock layer. At that point they segregate according to their density; gas is lighter so it will form a 'pool' immediately beneath the permeability barrier, whereas oil is heavier and will accumulate beneath the gas. Rocks beneath will be saturated with water or brine (Fig. 1.9).

Occasionally oil and gas are not stopped in their movement upward by geological structures that get in the way, and they find their way to the surface of the Earth. Examples of this include 'tar pits' like those of Trinidad, or the eternal flames that burn in the desert near Kirkuk in Iraq (Fig. 1.10).

Methane Hydrates

Methane hydrate deposits are a kind of shortcut fossil fuel in that they contain natural gas that mainly has not been buried very deep but which is still possible to extract from frozen soils or deep seabed sediments. Methane hydrates are ice-like substances which belong to a family of compounds known as clathrates, which comprise a lattice cage of a 'host' molecule within which 'guest molecules' are trapped. The hydrate 'cage' allows the methane molecules to be held close together, much closer than in a free gas, so small amounts of hydrate hold a lot of gas. Because of the size of

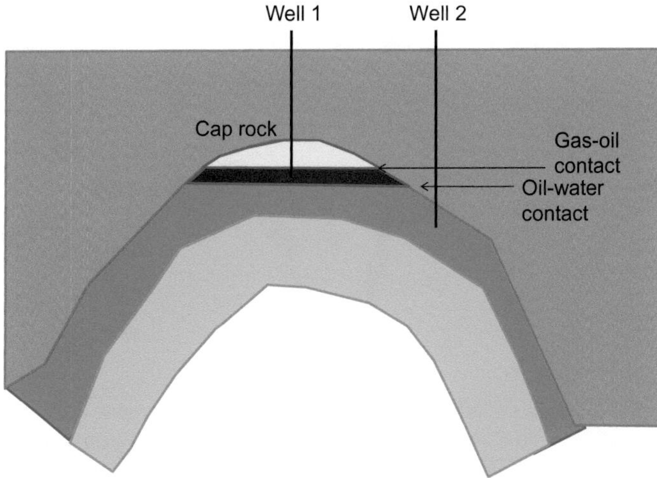

Fig. 1.9 Oil and gas under a caprock arch. Oil and gas migrate upwards until they are trapped beneath an impermeable rock layer. At that point they segregate according to their density; gas is lighter so it will form a 'pool' immediately beneath the permeability barrier, whereas oil is heavier and will accumulate beneath the gas. Rocks beneath will be saturated with water or brine.

Fig. 1.10 The 'eternal flames' near Kirkuk in Iraq. Here oil and gas are not stopped in their movement upward by geological structures, so find their way to the surface of the Earth. *(From https://upload.wikimedia.org/wikipedia/commons/7/7a/P3110004.jpg.)*

Fig. 1.11 Main areas of offshore methane hydrates. *(Redrawn from Klauda, J.B., Sandler, S.I., 2005. Global distribution of methane hydrate in ocean sediment. Energy Fuels 19, 469–470.)*

the area postulated for subsea methane hydrates, their global resource is considered very large, greater than for all other hydrocarbon energy sources (Fig. 1.11). Global methane hydrate resources are estimated at 2×10^{14} m^3 of methane. However, unlike other resources, for example conventional oil and gas or coal bed methane, their occurrence and relationship with host sediments is poorly understood, as are methods to extract them in deep-sea locations.

But where does the methane come from that is captured by the clathrate? In seabed sediment, methane can be generated by microbiological activity. It can also come from the deep rocks below (perhaps having seeped up from a conventional geological gas accumulation). It is possible to distinguish the deep and shallow gas isotopically and in most cases isotopic signatures suggest that naturally occurring subseabed hydrates are formed mainly from microbial methane, probably generated in situ by degradation of marine organic matter trapped in the sediment and then 'frozen' into clathrate.

Methane hydrates have received a lot of attention in parts of Asia, for example Japan and India, where large amounts of conventional geological natural gas are not available. However, commercial gas production from offshore methane hydrates has not been established anywhere in the world. In Japan, where extraction technology is advanced, a pilot test was established in March 2013 in the eastern Nankai Trough, which is believed to contain 20 trillion cubic feet of methane hydrate gas. One production well and two monitoring wells were drilled into sands at a depth of 300 m below seafloor and in 1000 m of seawater. In May 2017, the China Geological Survey reported successful extraction of methane hydrate gas in the Shenhu Sea, about 320 km southeast of Zhuhai City in Guangdong.

CARBONATE ROCKS

The formation of limestone is an important way that atmospheric CO_2 finds its way into rocks. Limestone is a sedimentary rock, made up of fragments of skeletons and shells of marine organisms such as coral, foraminifera, brachiopods and molluscs as well as the minerals calcite and aragonite, which are different crystal forms of calcium carbonate ($CaCO_3$). These elements all form in sea (and lake) water, and when they are permanently preserved in rock layers the CO_2 has been taken out of the atmosphere. About 10% of sedimentary rocks are limestones, so this is a significant way that CO_2 is sequestered; however, the process does not directly result in fossil fuels and so does not have a direct connection with energy.

Though the story of sequestration of carbon—its journey from the living world to the rocks—is clearly complicated, there is a simple principle at work here: that is, a huge exchange of carbon is going on all the time. This is the geological part of the carbon cycle, and part of it is what produced fossil fuels. The geological sequestration of carbon likely changes Earth's climate in the long term, as we will see in the next section.

But it is also the use of these fossil fuels that is the main cause of modern climate change in that humankind has caused a quick reversal of slow geological processes, putting some of that carbon back into the biosphere in the form of atmospheric CO_2.

HOW FOSSIL FUEL FORMATION AND COMBUSTION CHANGE CLIMATE

It is clear that fossil-fuel formation through the geological carbon cycle causes large-scale sequestration of carbon in the deep Earth. The use of that carbon in its many forms as fuel reverses the process. What is known about the details of these processes?

Geological Sequestration and Climate

The effects on ancient atmospheres of large-scale sequestration of carbon through coal, oil and gas, or methane hydrate formation are difficult to measure.

But if we take the formation of coal, some simple calculations can be made. The large coal-forming lycopsid trees of 310 million years ago that I mentioned earlier could have contained as much as 3200 kg of carbon per tree, and if this was buried in a subsiding basin and converted into coal, then most of the carbon would have been buried and isolated from the atmosphere. Two geologists, Chris Cleal and Barry Thomas, estimated that coal swamps grew from an area of 500,000 km^2 around 310 million years ago to over 2,000,000 km^2 10 million years later. Using a series of calculations assuming a certain lifespan for plants and density per square kilometre, they estimated that a stand of lycopsid trees might have captured and buried (through geological sedimentation and coalification) between 108 and 390 tonnes/hectare (t/ha) of carbon per year. Linking this figure with the maximum extent of the coal swamps gives a figure for the carbon capture rate of the Carboniferous coal swamps of almost 100 billion tonnes of carbon per year. Further calculations reveal that this would translate into a reduction of 44 parts per million (ppm) of CO_2 per year in the Carboniferous atmosphere. Cleal and Thomas are the first to admit that these figures are rather approximate, but when we compare the Carboniferous atmosphere to our own where human activities *add* a few parts CO_2 ppm per decade, it seems a very significant result.

Apart from the theory, is there any evidence that this huge level of carbon sequestration was happening 310 million years ago? Evidence from isotopes can help. CO_2 has two main carbon isotope components, ^{12}C and ^{13}C, and their ratio is expressed as $\delta^{13}C$. The two isotopes are processed differently by plant photosynthesis. Plants 'prefer' ^{12}C for their tissues but take in both isotopes, 'freezing' the ratio in their tissues. The $\delta^{13}C$ is therefore a record of the atmosphere at the time that the plants were growing. But also when plant growth is associated with large amounts of carbon burial—as in the Carboniferous coal swamps—the burial alters the balance of $\delta^{13}C$ in the atmosphere of the time. Lots of burial of carbon causes atmospheric $\delta^{13}C$ to increase and these increases are passed on to new plants. This means that $\delta^{13}C$ of plant tissue through time records increases in carbon burial, for example rapid expansion of swamp forests. Conversely, a decline in the swamp forests would bring about a decrease in $\delta^{13}C$.

The record of $\delta^{13}C$ in fossil wood between about 430 and 250 million years ago (Fig. 1.12) shows several trends. The increase of $\delta^{13}C$ in the early stages of the Carboniferous about 360 million years ago was probably caused by the carbon burial associated with the expansion of plants across the previously barren land, and the formation of limestone. The period between

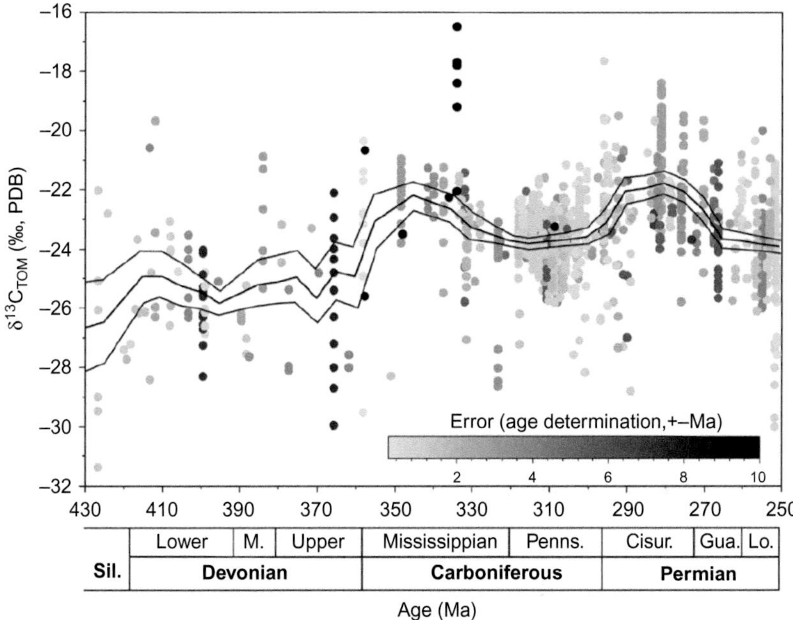

Fig. 1.12 δ^{13}C of fossil wood between 430 and 250 million years ago. *(From Peters-Kottig, W., Strauss, H., Kerp, H., 2006. The land plant 13C record and plant evolution in the Late Palaeozoic. Palaeogeogr. Palaeoclimatol. Palaeoecol. 240, 237–252.)*

330 and 300 million years ago, over which the coal swamps grew, is associated with a gentle rise in δ^{13}C.

Transfer of carbon from the biosphere to the geosphere does not only occur through burial of plant material as coal, or through the formation of limestone. Another large-scale geological process that does the same thing is weathering of silicate minerals which convert atmospheric CO_2 mainly to soluble bicarbonate ions, which flow in rivers to the sea and form a reservoir for solid carbonate secreting animals. Because silicate minerals are what make up much of the hard rock of the Earth, and particularly mountain chains, the creation and thrusting up of mountain chains is often associated with carbon capture because so much rock is newly exposed to fast weathering.

As I mentioned earlier, the Variscan Ranges were being created around the time of the coal swamps and weathering would have been particularly intense, as they were born close to the position of the Carboniferous equator. This occurring at about the same time as the expansion of the coal swamps would have created a powerful 'sink' for atmospheric CO_2.

Perhaps the coal swamps and the simultaneous creation of the Variscan Ranges lead to a 'negative greenhouse' condition in the later part of the Carboniferous—basically the reverse of what we see now with increasing levels of CO_2 in the atmosphere. Though the evidence of atmospheric change from $\delta^{13}C$ is not particularly clear, there is abundant evidence for a glacial period, in fact one of the longest and geographically widespread in geological time—sometimes known as the Permo-Carboniferous glaciation—which reached its greatest intensity around 300 million years ago (Fig. 1.13).

What are the climate-changing effects of large-scale marine sequestration of the type that initiates the formation of oil and gas? An interesting example of this comes from even further back in geological time than the Carboniferous—around the end of the Ordovician and the beginning of the Silurian period about 444 million years ago. The atmosphere and land at that time was very different; for example, atmospheric CO_2 levels were very high with the result that much of the Ordovician and Silurian were 'greenhouse' periods. Long before the evolution of land plants most of the organic carbon reservoir must have existed in the oceans, because carbonate burial was relatively scarce.

Fig. 1.13 Evidence of the Permo-Carboniferous glacial period about 300 million years ago that affected the southern continents. Left—scratches on an ancient rock surface left by glaciers. Right—a 'dropstone' dropped from floating ice into glacial lake sediments. Both from the Oman Huqf area. *(Photos M.H. Stephenson.)*

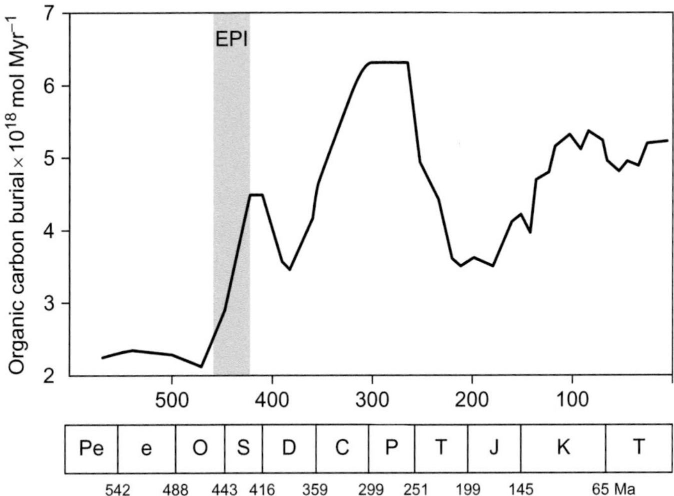

Fig. 1.14 Temporal changes in organic carbon burial flux, showing a notable high in the Early Palaeozoic Icehouse (EPI) when organic carbon was predominantly buried in black shales. *(From Page, A., et al., 2007. Were transgressive black shales a negative feedback modulating glacioeustasy in the Early Palaeozoic Icehouse? In: Deep Time Perspectives on Climate Change. The Geological Society, London, pp. 123–156.)*

Computer modelling using the carbon isotopic composition of seawater (Fig. 1.14) indicates theoretical rates of carbon burial varying through geological time with high rates of burial 440 and 320–280 million years ago. The latter episode corresponds to the creation of the Carboniferous coal swamps. An earlier episode of burial corresponds closely to the deposition of large volumes of black shale, an organic-rich, fine-grained, dark-coloured sedimentary rock, essentially originating from deposition in deep anoxic seas.

Despite the high CO_2 of the atmosphere, evidence from the rocks like those in Fig. 1.14 indicates a period of glaciation around 445 million years ago variously called the Early Palaeozoic Icehouse and the 'Hirnantian Glaciation'.

A team of geologists led by Alex Page suggested that the deposition of black shales was sufficient to create the conditions for cooling and glaciation. They compared the thicknesses and ages of black shale layers with sea level curves (influenced by the rise and fall of glaciers), isotopic data, and direct rock evidence of ancient glaciation to establish a connection. They also speculated that the reason why 'runaway' global warming did not occur even with high atmospheric CO_2 was because of the moderating influence of black shale CO_2 'drawdown'. Interestingly, Page also considered that melting ice in greenhouse conditions would trigger higher sea level, more anoxic oceans and therefore a brake on higher temperature, in other words a negative feedback.

This idea that concentrated periods of intense geological sequestration take place after major glacial periods due to sea level rise introduces an interesting twist to the story. According to the German geologist Sebastian Lüning, following the Hirnantian glacial period, ice melted and sea level rose, flooding glaciated lowland areas across what is now North Africa and the Middle East. The new shallow seas were fed by nutrients from the land and as the water warmed up they teemed with life. The hollows of the flooded landscape created low-oxygen havens for the preservation of organic matter. The result was a rock with very high organic content that today sits far below the surface across North Africa and the Middle East. Following heat and pressure, the rock matured and is now the source of most of the region's older oil and gas deposits, as well as the source of gas of the Qatari South Pars gas field, which ironically itself supplies much of Britain's liquefied natural gas (LNG). This is the gas that Britain and other countries that import Qatari LNG burn increasingly in power stations.

Combustion and Climate Change

In a book simply about climate change, the process by which combustion of fossil fuels produces climate change would take centre stage. But here the combustion is seen simply as part of the process that returns buried carbon back to the atmosphere. We have seen earlier that this process can happen naturally, for example in the eternal flames near Kirkuk in Iraq, or on a large scale when coal deposits are eroded and begin to oxidise in the atmosphere, but the central problem of modern climate change is that human activity accelerates this process hugely. Largely because of the industrial revolution, coal was mined and burned systematically, followed swiftly by oil, putting enormous amounts of CO_2 in the atmosphere.

But why does having CO_2 in the atmosphere cause global warming? There are two components in the greenhouse effect—a natural one caused by the amounts of greenhouse gases naturally found in the atmosphere—and a man-made one caused by the greenhouse gases that human activities add to the atmosphere. The contribution that a greenhouse gas makes to the greenhouse effect depends on the amount of it in the atmosphere and how much heat that particular gas absorbs and reradiates. In descending order, the gases that contribute most to the Earth's greenhouse effect are: water vapour, carbon dioxide, nitrous oxide (N_2O), methane (CH_4) and ozone (O_3), but this is mainly due to the relative amounts of these gases. Their effectiveness as a greenhouse gas apart from the abundance is known as their global warming potential, or GWP. In these terms, methane is 23 times more effective, and nitrous oxide is 296 times more effective, than carbon dioxide at warming the climate.

Not all the greenhouse gas emitted to the atmosphere stays there indefinitely. For example, the amount of CO_2 in the atmosphere and the amount of CO_2 dissolved in surface waters of the oceans stay in equilibrium, because the air and water mix well at the sea surface. Thus when we add more CO_2 to the atmosphere, a proportion of it dissolves in the ocean.

If there was no natural greenhouse effect, the heat emitted by the Earth would simply pass outwards from the Earth's surface into space. So without it, the Earth's average surface temperature of 14°C would be as low as −18°C, the so-called 'black body' temperature of the Earth. In our solar system, Mars, Venus, and Saturn's moon Titan have surface and lower atmospheric temperatures that are consistent with the predicted greenhouse effects of their atmospheres.

SUMMARY

This chapter has tried to show the intimate links that the carbon cycle bestows to fossil fuels and climate change. In effect the two are bound together. The long-term geological carbon cycle takes large amounts of carbon from the atmosphere and the biosphere and sequesters it in deep isolated rocks—and if it does not emerge, it is essentially out of harm's way. In the process, sequestration probably mostly cools the Earth. But of course this sequestered carbon does emerge in volcanoes and in the weathering of carbonate rocks (for example). It also emerges because humankind burns the carbon of fossil fuel, a sort of free concentrated form of ancient energy. Later in the book I will examine how what we know about geological history helps to identify some of the big events of the carbon cycle, including mass extinctions of life, and the anatomy of humankind's heretofore smaller interventions in the cycle through the 'fossil economy'.

BIBLIOGRAPHY

Beerling, D., 2007. The Emerald Planet: How Plants Changed Earth's History. Oxford University Press. 416pp.

Berner, R., 2003. The long-term carbon cycle, fossil fuels and atmospheric composition. Nature 426, 323–326.

Cleal, C.J., Thomas, B.A., 2005. Palaeozoic tropical rainforests and their effect on global climates: is the past the key to the present? Geobiology 3, 13–31.

Cleal, C.J., et al., 2010. Late Moscovian terrestrial biotas and palaeoenvironments of Variscan Euramerica. Netherlands J. Geosci. 88, 181–278.

International Energy Agency 2016. Key World Energy Statistics. https://www.iea.org/publications/freepublications/publication/KeyWorld2016.pdf.

Klauda, J.B., Sandler, S.I., 2005. Global distribution of methane hydrate in ocean sediment. Energy Fuels 19, 469–470.

Lenton, T., 2016. Earth System Science: A Very Short Introduction. Oxford University Press. 176pp.

Letcher, T. (Ed.), 2013. Future Energy: Improved, Sustainable and Clean Options for our Planet, second ed. Elsevier Science. 738pp.

Link, P., 2007. Basic Petroleum Geology, third ed. Pennwell. 443pp.

Lüning, S., et al., 2000. Lowermost Silurian 'Hot Shales' in North Africa and Arabia: regional distribution and depositional model. Earth Sci. Rev. 49, 121–200.

McCabe, P.S., 1987. Facies studies of coal and coal-bearing strata recent advances. Special Publications of the Geological Society 32, 51–66.

Page, A., et al., 2007. Were transgressive black shales a negative feedback modulating glacioeustasy in the Early Palaeozoic Icehouse? In: Deep Time Perspectives on Climate Change. The Geological Society, London, pp. 123–156.

Peters-Kottig, W., Strauss, H., Kerp, H., 2006. The land plant ^{13}C record and plant evolution in the Late Palaeozoic. Palaeogeogr. Palaeoclimatol. Palaeoecol. 240, 237–252.

Prud'homme, A., 2014. Hydrofracking: what everyone needs to know. Oxford Univ Press, Oxford, UK. 208pp.

Stephenson, M.H., 2014. Five unconventional fuels: geology and environment. In: Unconventional Fossil Fuels: The Next Hydrocarbon Revolution? Emirates Center for Strategic Studies and Research, Abu Dhabi, pp. 13–34.

Stephenson, M.H., et al., 2010. Northern England Serpukhovian (early Namurian) farfield responses to southern hemisphere glaciation. J. Geol. Soc. 167, 1171–1184.

Strauss, H., Peters-Kottig, W., 2003. The Paleozoic to Mesozoic carbon cycle revisited: the carbon isotopic composition of terrestrial organic matter. Geochem. Geophys. Geosyst. 4, 1–15.

Wang, Q., Chen, X., Jha, A.N., Rogers, H., 2014. Natural gas from shale formation—the evolution, evidences and challenges of shale gas revolution in United States. Renew. Sustain. Energy Rev. 30, 1–28.

Wignall, P.B., 1994. Black Shales (Oxford Monographs on Geology and Geophysics). Oxford University Press. 144pp.

CHAPTER 2

Natural Global Warming: Climate Change in 'Deep Time'

Contents

In this chapter I will consider the footprint that the geological part of the carbon cycle has had on the deep history of the Earth. Sedimentary rocks and unconsolidated sediments contain preserved elements of the ancient environments in which they were deposited through their sedimentary structures, fossils, isotopic and geochemical character. The layers of sedimentary rocks can essentially be 'read' and interpreted to identify cycles of warming and cooling spanning huge periods of millions of years. A few of the warming episodes have been severe enough to affect the continuity of life on Earth: for example at the Permian-Triassic boundary about 250 million years ago and at the Palaeocene-Eocene Thermal Maximum about 55 million years ago. These geological studies enable us to see climate change being acted out on a grand scale that simply is not visible to modern climate scientists. Its value is that you can see the broad sweep of change, or 'the wood for the trees'. But deep-time climate change information has little value for the planners and policymakers that seek the short-term detailed and local information that can futureproof our society to climate change.

One phrase that historians like is Edmund Burke's 'Those who don't know history are destined to repeat it.' The theme of this chapter is the same, though geological history goes back a lot further than human history. What can we learn about the long-term workings of the carbon cycle from the thousands of millions of years of geological history? The conditions of the first few thousand million years were so different to our modern world that not much can be learned. However, as the Earth entered the last few hundred million years before the present, there were several cycles of warming and cooling that can tell us something about how the Earth changed and

Energy and Climate Change
https://doi.org/10.1016/B978-0-12-812021-7.00002-6

perhaps what might be in store for us in the near future. But how do mere rocks tell us all this?

Palaeontologists and stratigraphers (the latter are geologists that study the layers or strata of rocks) look on rocks quite differently to the way that (for example) petroleum geologists do. Petroleum geologists see the properties of rocks and how they might act as containers or conduits for useful fluids like gas or oil. Palaeontologists and stratigraphers look at rocks as 'recorders' of changes in the ancient environment in which the rocks formed. The thin layers of shale, for example—a fine-grained dark-coloured sedimentary rock—are able to record aspects of the environment in which they formed. The most obvious thing they tell us is the kind of sediment that was deposited, but they also contain fossils of once-living organisms and chemical traces of the ancient environment. Oxygen isotopes provide information on temperature and ice volume; carbon isotopes reflect global biomass and inputs (of methane or carbon dioxide) into the ocean and atmosphere; and strontium and osmium are proxies for weathering.

Mostly we think of fossils as faint impressions of shells or ammonites on the surfaces of rock layers, but fossils can also be very small—too small to see—for example, fossil spores or pollen, or fossil plankton. Rocks sometimes contain thousands of fossils of this type in a single thin fragment.

Taking shale again, each thin layer represents a single episode of sedimentation on the seabed or a lakebed. Breaking open the shale, splitting it along its layers, exposes surfaces that were once part of an ancient seabed. In fact the layers of shale seen side-on sometimes look like the pages of a book and those pages can be 'read' or interpreted like a book. The 'pages' provide information on how the environment changed through time.

So for palaeontologists or stratigraphers an outcrop of layered sedimentary rock can be an insight into a period of ancient environmental change. These scientists will measure the outcrop in great detail, noting changes in the grain size of the sediment, the fossil content, the organic matter content, and the chemistry of the rock. They will likely try to interpret what they see, and perhaps look for some corroboration of the sequences of events that they interpret, by looking at other outcrops. Cores of rock recovered from boreholes can be used in the same way. Most of what we know about the geological history of our planet results from this kind of study, and a generalised history of climate over the last 3.8 billion years of 'deep time' can be constructed.

Fig. 2.1 illustrates atmospheric temperature variation at six different timescales (A) showing most of the whole of geological history to present, (B) showing the last 600 million years, and (C) the last 70 million years. (D) to (F) show smaller periods of time closer and closer to present. In the temperature variation shown, some of the global events that

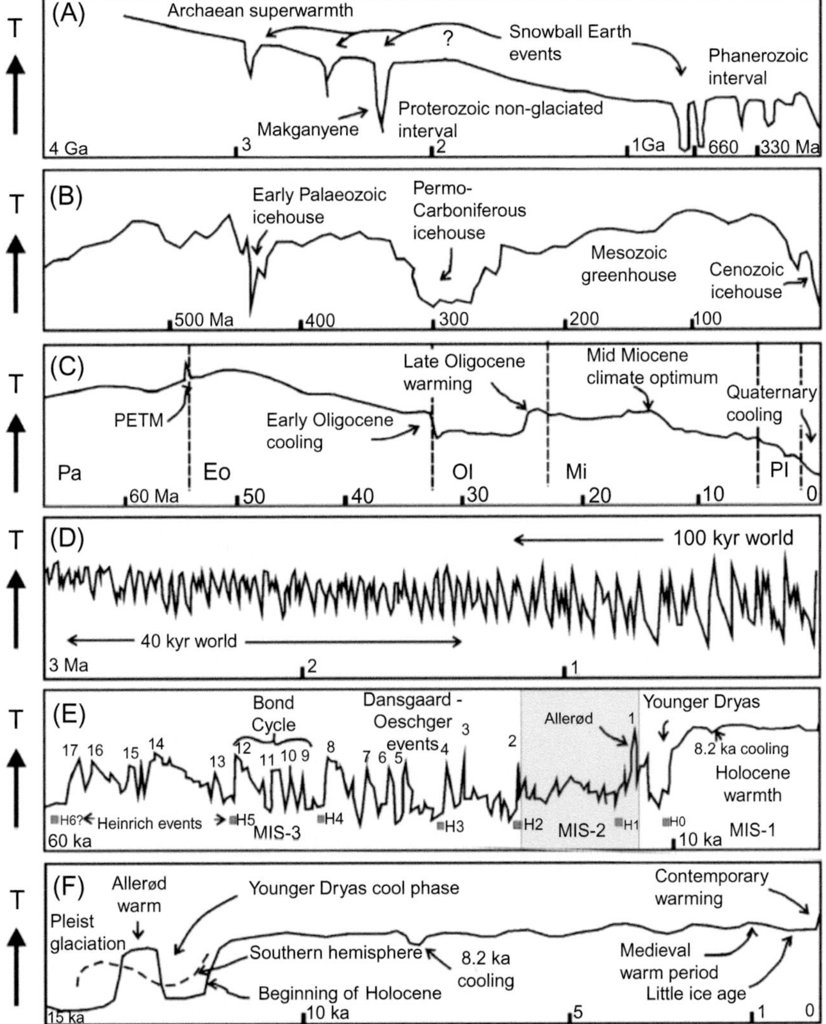

Fig. 2.1 Global climate variation at six different timescales. 'T' indicates relative temperature. *(From Zalasiewicz, J., & Williams, M. (2016). Climate change through Earth's history. In Trevor M. Letcher,* Climate change *(pp. 3–17), second ed. Elsevier.)*

I examine in the next section can be seen, for example the Palaeocene-Eocene Thermal Maximum (PETM) and the Younger Dryas.

THE PALAEOCENE-EOCENE THERMAL MAXIMUM

For the purposes of this book the most important events are periods of global warming in deep time. Perhaps the best known of these is the PETM, which shows how Earth can warm up quickly and how life in the oceans and on land can die back, but also how life can recover.

Most studies of the PETM begin with a close examination of a rock borehole core, which is a cylinder of rock taken from a drilling rig. The core is often sliced lengthways and divided up so that the rock can be examined along its length. A slice of a core from a borehole drilled in the central part of the North Sea, which passes through the PETM, is shown in Fig. 2.2.

Fig. 2.2 Rock core from the Palaeocene-Eocene Thermal Maximum (PETM). The core has been sliced vertically to create a slab. *(Photo: M Stephenson.)*

The dark-coloured rock is shale and the number markings are depths below ground level, in this case between 8594 and 8600 ft (2619.5 m and 2621.3 m). Most of the biggest changes of the PETM took place between about 2621 and 2609 m in this borehole core, so the part of the core in the picture shows the rock just at the start of those changes.

Having said that, the rock itself does not appear to alter much through the big changes of the PETM. A simple visual examination tells us very little. But sophisticated study reveals many changes (Fig. 2.3).

The diagram shows the depth in the borehole core (shown in the photo in Fig. 2.2) on the left, increasing downward; the lower the level in the borehole core, the older the rock is. Recording the fossils and chemistry of the rock in the borehole going *upward* would mean following the progress of

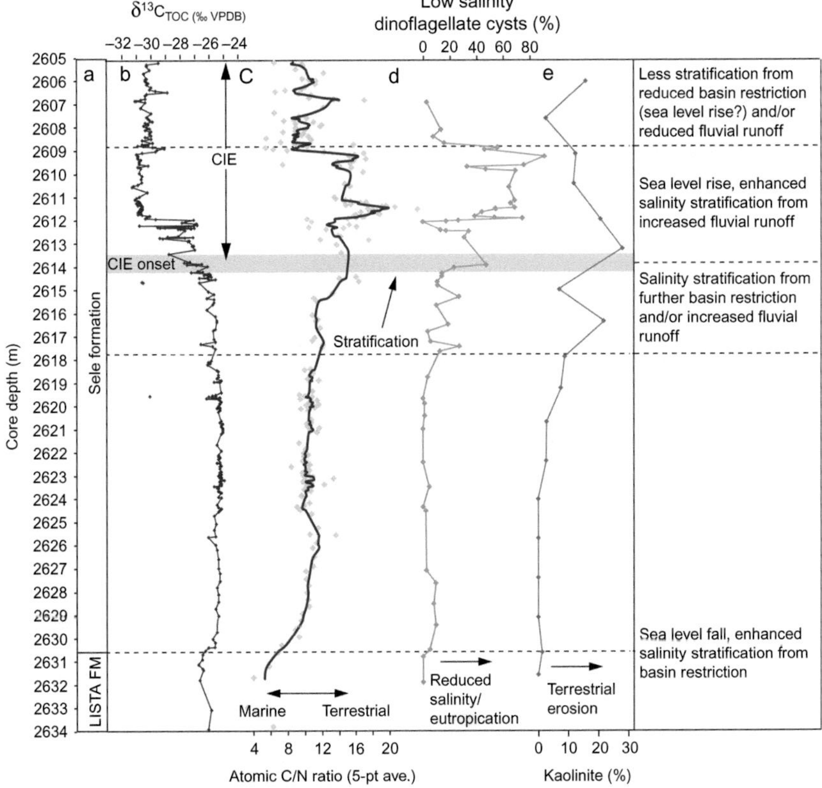

Fig. 2.3 Analyses of fossil and geochemical content of the PETM in the rock core from beneath the North Sea. *(From Kender, S. et al., 2012. Enhanced precipitation and vegetation changes in the North-East Atlantic at the Palaeocene-Eocene boundary. Earth Planet. Sci. Lett. 353–354, 108–120.)*

time, seeing the changes in the order they occurred. Amongst the measurements that were made on this piece of borehole core were carbon to nitrogen ratio, the percentage of the mineral kaolinite in the shale, the $\delta^{13}C$ of the organic carbon in the shale, and the number of fossil 'low-salinity' dinoflagellate cysts counted. These measurements and others allowed an interpretation on the right-hand side of the diagram.

But the part to concentrate on is on the left. The $\delta^{13}C$ of organic carbon in the shale has an extraordinarily large jump to the left between 2615 and 2612 m, which suggests that the atmosphere and the carbon cycle must have been changing a lot—and very rapidly—in this part of the ancient North Sea.

Elsewhere rock cores of the PETM have revealed fossil shells that allow us to see other changes in isotopic ratio. For example, oxygen isotopes through the time of the PETM tell us something about the temperature of seawater. The startling results are that for a very short period of Earth's history (less than 200,000 years), sea surface temperatures rose by between 5 and 8°C, but then decreased again. This spike in temperature is shown in Fig. 2.4 (top left of the graph) and the 200,000-year period is so thin that it looks like a single line on the scale of million-year time intervals. The temperature graph suggests that the Earth's oceans reached their warmest for 65 million years during this short time period.

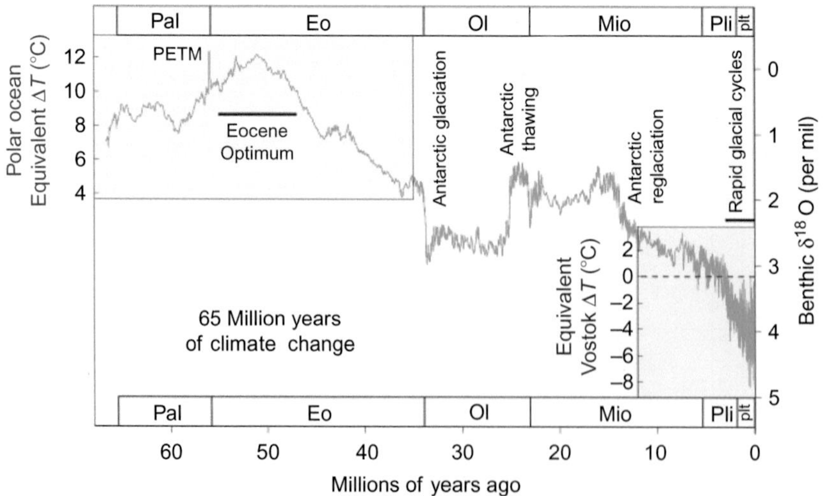

Fig. 2.4 The temperature spike (in the green line) of the PETM from $\delta^{18}O$ of fossil shells. *(From Wikipedia, retrieved July 2017.)*

It is known that the sea became more acidic during the PETM and that there were extinctions in the Earth's seas, either because of the acidity or the warmer temperatures. Many ocean plankton, for example, simply could not cope with the changes. The sea level rose as well because of ice melting, but also perhaps because of the thermal expansion of seawater. The effects on land are uncertain because the PETM is best represented in marine sediments, but there is some indirect evidence.

The North Sea, for example, looked rather different during the PETM (Fig. 2.5). It was more enclosed by land, except to the north. So there was probably no English Channel at the time. The Atlantic Ocean was much narrower and Iceland did not exist yet.

Because the North Sea was more enclosed, the sediments deposited during the PETM contain more information about the changes on the land. Fig. 2.3 shows percentage levels of 'low-salinity dinoflagellate cysts'. Dinoflagellate cysts are the fossils of a particular kind of plankton and the ones that were measured for the graph were those that we know preferred low-salinity seawater. What their increase seems to suggest is that the surface waters of the North Sea became less saline during the PETM. This might have been because there was suddenly a lot more rainfall on the land around and this was brought by rivers, making the sea less saline. But to cause such a huge salinity change in a large sea implies an enormous input of fresh water from the land. The spores and pollen from land plants around the North Sea which we find in the sediment layers at this point also suggest whole ecologies substituting one for another as the sea level and temperature rose. These are just a few of the environmental changes that have been interpreted from the data. We are not sure of the effect this would have had on wider ecology or on a society like our own, if it had existed at the time. But it is likely that the changes were caused directly or indirectly by global warming.

How does the rate of change through the PETM compare with modern forecasts of global warming? The most ambitious of the plans for emissions reduction from the Intergovernmental Panel on Climate Change (IPCC) want to ensure a limit of temperature increase to 2°C in the next century. Recent work on the PETM suggests seawater temperature increase from oxygen isotopes and other chemical methods over about 20,000 years was between 5°C and 8°C. The world of the PETM was essentially ice-free. What this suggests then is that the *present* rate of change in seawater is *already* much faster than seems to have happened over the PETM. Temperature did

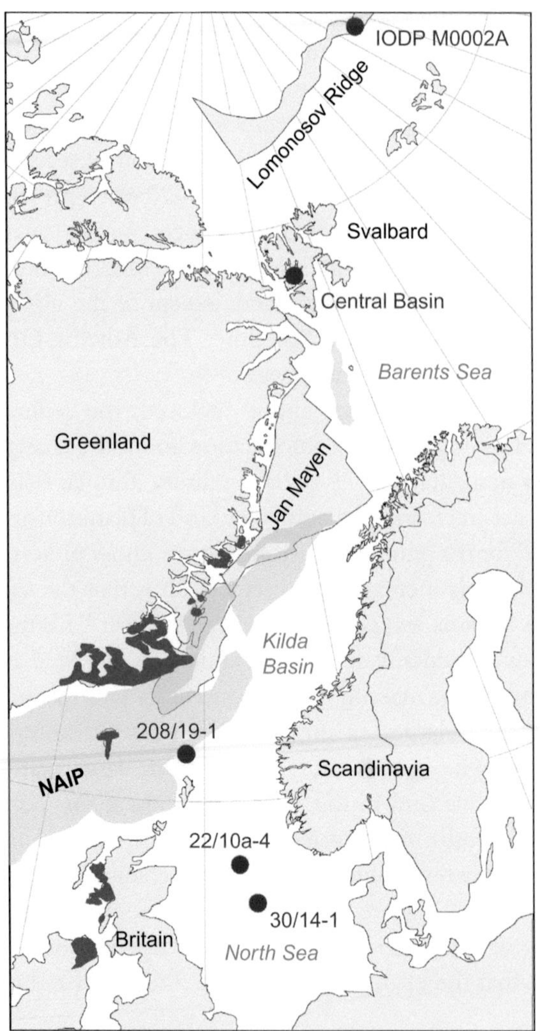

Fig. 2.5 The North Sea was more enclosed by land 55 million years ago, during the time of the PETM. *(From Kender, S. et al., 2012. Enhanced precipitation and vegetation changes in the North-East Atlantic at the Palaeocene-Eocene boundary. Earth Planet. Sci. Lett. 353–354, 108–120.)*

drop after the PETM but this took between 30,000 and 150,000 years. Big natural global carbon storage processes probably rectified things, but not fast enough to save quite a few species.

What caused the PETM? In the absence of organised industrial activity 55 million years ago, the cause and changes were clearly natural. It seems likely that greenhouse gases (either CO_2 or methane) were being added to the atmosphere at a very rapid rate, rather like they are now. The size of the change in $\delta^{13}C$ in the PETM atmosphere is such that only very large releases of CO_2 or of methane with very low $\delta^{13}C$ can explain it.

Volcanoes can produce enormous amounts of CO_2 in eruptions, but in order to explain the $\delta^{13}C$ change, around 1500 gigatons (GT) of carbon would have to have been emitted from volcanoes over the period of the PETM. This is many times more than normal, even when volcanoes have been very active in deep time.

The most favoured explanation at the moment is that methane was released suddenly from methane hydrates in sea-bottom sediments. These were discussed in the last section. Methane of this type has very low $\delta^{13}C$, and as a very potent greenhouse gas methane has great global warming potential. Methane can be released from methane hydrates following warming and/or pressure release.

Methane in methane hydrate is not normally seen as a fossil fuel, partly because its use so far as a fuel is very limited, and partly because in most cases the methane is not primarily of geological but of biological origin; however, its presence in methane hydrate at the sea floor is still an example of the long-term geological carbon cycle at work.

The fact that methane from methane hydrate could become a fuel—for example in Japan—and that its ancient release may have been a cause of the PETM is a good illustration of fossil fuel and its connection to rapid and catastrophic climate change.

THE PERMIAN-TRIASSIC EXTINCTION

The Permian-Triassic extinction, also known as the End Permian Extinction or more dramatically as the 'Great Dying', happened about 252 million years ago. It is the Earth's most severe known extinction event, with up to 96% of all marine species and 70% of terrestrial vertebrate species becoming extinct. It is probable that this extinction was related to high levels of greenhouse gases in the atmosphere, and more closely to global warming, acid rain, increased erosion on land, depletion of the oxygen of ocean water—and possibly to release of methane from hydrates. These changes led, on land and at sea, to changes in habitats that animals and plants could not adapt to, with the inevitable result of extinction.

Like the PETM, the 'Great Dying' can be located in layers of rock. One of the advantages of the fact that thin layers of volcanic ash occur above and below the level at which extinction occurred is that the age of the extinction can be calculated quite accurately within a range. This is because there is a technique for determining the age of layers of volcanic ash. The extinction can be bracketed in time between 251.941 ± 0.037 and 251.880 ± 0.031 million years ago, a period of around 60,000 years. As during the PETM, a big change in the isotopic composition of the atmosphere seems to have occurred as well as an 8°C rise in temperature. Detailed study has shown not one but three separate extinctions which—geologically speaking at least—were very close together in time (Fig. 2.6).

One of the best places to see the layers of rock which record the extinctions is at a quarry near Meishan in China. The quarry was once used for limestone for building, but in recent years has been recognised as a good place to study the End Permian Extinction, and also as an official location for the geological boundary between the Permian and Triassic periods. This boundary is very precisely located as the surface between 'layer 27b' and 'layer 27c' of a series of numbered and intensely studied rock layers (Fig. 2.7). It is not particularly important for our purposes, but the extinctions are.

The extinctions can also be quite precisely located at the quarry. It is important to realise that not all the extinctions of animals and plants can be seen at these levels in the rocks at Meishan; you have to visit other locations where the same layers outcrop to see the full picture. Chinese palaeontologists, particularly, have been involved in careful studies to illustrate just how much of the fauna and flora died out.

Perhaps because 252 million years is so far back, the reasons for the extinctions are difficult to divine. The most widely accepted explanation starts with the eruption of the volcanic Siberian flood basalts. The remains of these volcanic basalt rocks can be seen today over large areas of northern Siberia; but for about 600,000 years, 252 million years ago, these volcanoes generated large volumes of sulphate aerosols, carbon dioxide and methane. It is also possible that the volcanoes heated and oxidised ancient coal deposits so that further CO_2 was released.

A summary of what happened next (Fig. 2.8) speculates that sulphur dioxide, CO_2, and volcanic methane all played parts in causing acid rain, global warming, and increased continental weathering with obvious deleterious effects. Perhaps most intriguing is the part that continental weathering might have played. The story goes that increased weathering meant that there were much greater amounts of nutrients in coastal waters, supplied by rivers newly charged with sediment. This led to eutrophication or, more correctly,

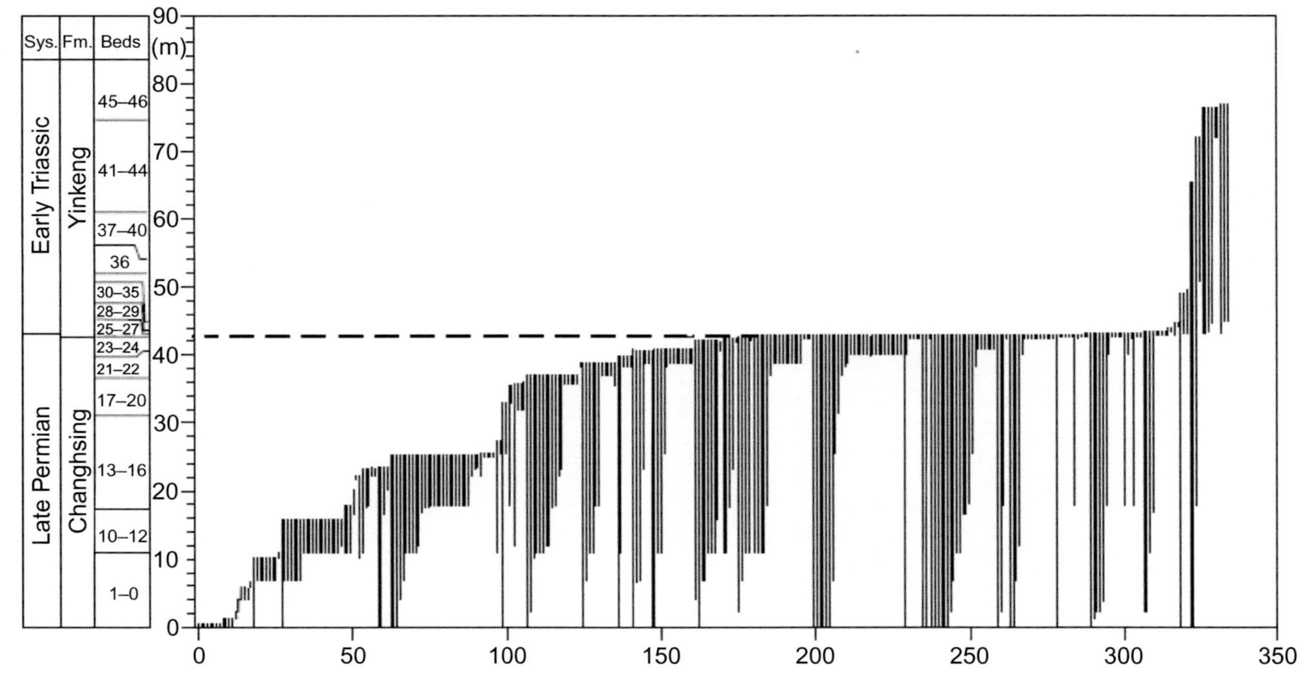

Fig. 2.6 Extinctions of fossil species at or near the Permian-Triassic boundary. The vertical lines indicate the ranges of fossils through layers of sedimentary rocks. A marked series of extinctions is clear around beds 24–27. *(From Jin, Y.G. et al., 2000. Pattern of marine mass extinction near the Permian-Triassic boundary in South China. Science 289, 432–436.)*

Fig. 2.7 The location in Meishan, China of the 'official' boundary between the Permian and Triassic periods with the levels of extinction marked in red. Extinction 2 was the most severe of the three extinctions. There was 85% species loss between layers 24e and 28 as shown on the slab beside the rock outcrop. The official geological boundary between the Permian and Triassic is the line between layers 27b and 27c. *(Photo M. Stephenson.)*

hypertrophication. This is the depletion of oxygen in a water body, a lake or a coastal area, which kills aquatic animals through lack of oxygen. It is a response to the addition of excess nutrients, mainly phosphates, which induces explosive growth of plants and algae, the decaying of which consumes oxygen from the water. This makes the sea anoxic.

It is clear that the extinction affected marine organisms with calcium carbonate skeletons (e.g., shells), probably because of ocean acidification from increased atmospheric CO_2.

THE CRETACEOUS–TERTIARY EXTINCTION

This mass extinction was not quite so severe as that at the Permian–Triassic boundary but nevertheless about 75% of the plant and animal species on Earth died out over a geologically short period of time, approximately 66 million years ago. The most famous of the animals that died out were the non-avian dinosaurs.

Like the other geological events I describe in this chapter, the evidence for the environmental change is contained within rocks that have been studied in great detail for many years. The work began with the identification of

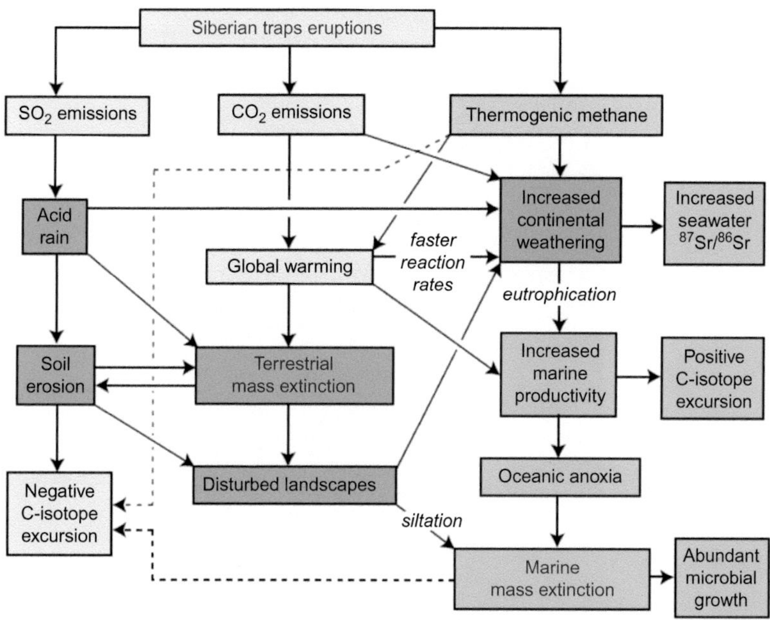

Fig. 2.8 Summary of some of the possible events following the Siberian Traps eruptions. *(From Benton, M.J., Newell, A.J., 2014. Impacts of global warming on Permo-Triassic terrestrial ecosystems. Gondwana Res. 25, 1308–1337.)*

a thin (1-cm thick) soft clay layer at the Cretaceous-Tertiary boundary in the Bottaccione Gorge at Gubbio in Central Italy. It was known at the time that extinction was characteristic of the Cretaceous-Tertiary boundary and locally in the rocks in central Italy. The Nobel Prize-winning experimental physicist Luis Alvarez used samples from the Bottaccione clay layer and above and below it, around Gubbio, to show that an iridium anomaly was present in the clay layer (Fig. 2.9). The original diagram from the paper published in the journal *Science* is reproduced here (Fig. 2.10). In the paper, Alvarez argued that the iridium was extraterrestrial and that it came from an asteroid impact.

The effects of such an impact would be a persistent 'winter' in which species perished, including on land non-avian dinosaurs, many mammals, pterosaurs, birds, lizards, insects and plants, and at sea plesiosaurs, giant marine lizards, many fish, sharks, molluscs, and many species of plankton.

Alvarez's idea seems to be supported by the discovery of the 180-km-wide Chicxulub Crater in the Gulf of Mexico in the early 1990s, perhaps caused by the asteroid; but some experts dispute the role of the impact and regard volcanic eruptions from the Indian Deccan Traps as a possible cause.

Fig. 2.9 The Cretaceous–Tertiary boundary in the Bottaccione Gorge at Gubbio in Central Italy. The geologist has his right hand on the clay layer. *(From http://www.liber-lapidum.net/imgitaly.html.)*

The Deccan Traps could have caused extinction through several mechanisms, including the release of dust and sulphuric aerosols into the air, which might have blocked sunlight and thereby reduced photosynthesis in plants. In addition, Deccan Trap volcanism might have resulted in carbon dioxide emissions that increased the greenhouse effect when the dust and aerosols cleared from the atmosphere.

Without going into detail, it is clear that rapid environmental change occurred. As with the Permian-Triassic boundary, it is possible that rising greenhouse gas and global warming may have had a role in extinction.

THE END OF THE YOUNGER DRYAS

After the PETM, the Younger Dryas (named after the cold climate plant *Dryas octopetala*) is perhaps the best-known example of deep-time rapid climatic change. Beginning about 14,500 years ago, the climate began to slowly warm up, but then there was a sudden lapse back into cold conditions followed by a very rapid warming 11,500 years ago. The stages of this warming are not, this time, shown in rock layers but in ice cores.

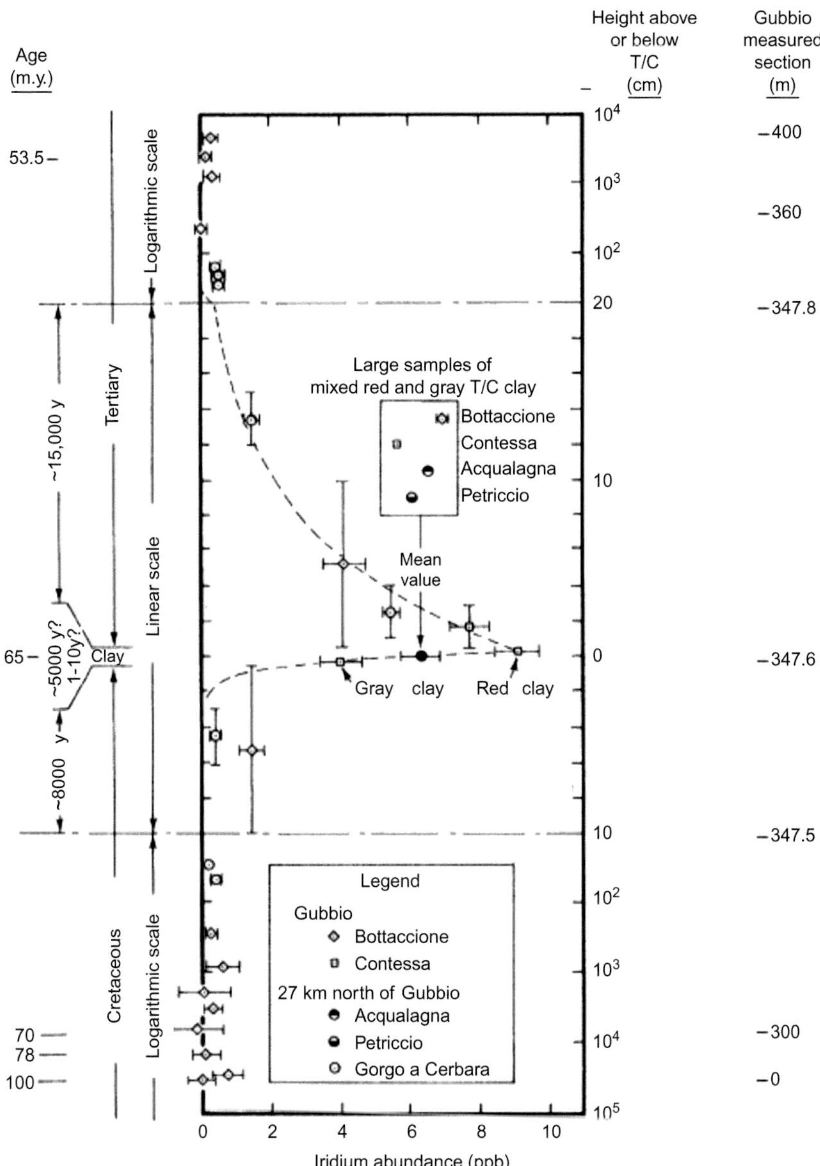

Fig. 2.10 The iridium anomaly at the Cretaceous-Tertiary boundary at Bottaccione, Central Italy. Iridium peaks in the thin clay layer. *(From Alvarez, L.W. et al., 1980. Extraterrestrial cause for the Cretaceous-Tertiary extinction. Science 208, 1095–1108.)*

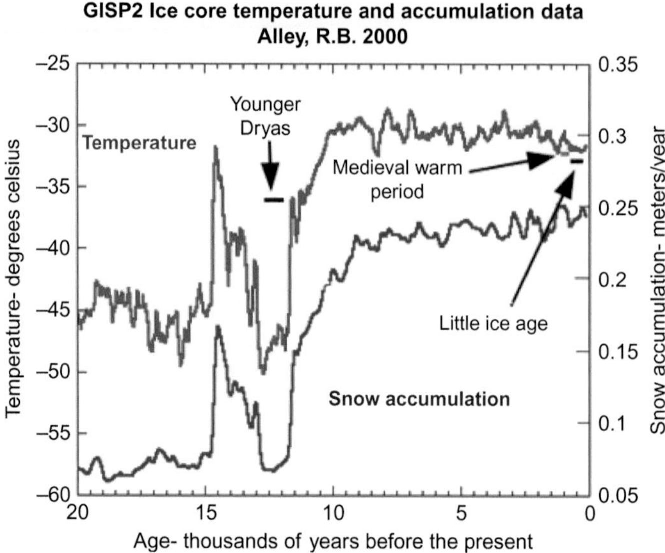

Fig. 2.11 The temperature change at the Younger Dryas. *(From https://www.ncdc.noaa.gov/paleo/pubs/alley2000/alley2000.html.)*

A particular borehole, the GISP2 (Greenland Ice Sheet Project Two), is important. This borehole was drilled through 3053.44 m of ice in central Greenland and a core was recovered through the whole length. Like sedimentary rock, ice collects in layers which record aspects of the environment including, through oxygen isotopes, the temperature of the time at which the ice formed. The result of some of this analysis (Fig. 2.11) shows the position of the Younger Dryas.

The Younger Dryas is also visible in other places. In the Cariaco Basin in Venezuela, temperatures dropped by 3°C. In parts of the Northern Hemisphere tropics, it became drier. In Antarctica it is somewhat different, however. Ice cores there record temperature rising during the Younger Dryas.

Despite being the latest of the examples I have chosen of rapid climate change, the Younger Dryas is one of the hardest to explain. It was a sudden cooling during an otherwise slow transition from the last glacial period into the present interglacial period, known as the Holocene. At the start of this slow warming, ice sheets were melting. Climate models show that this would have added fresh meltwater to the North Atlantic. At the start of the Younger Dryas, it has been shown that meltwater flows into the North Atlantic increased. Geologists think that fresh meltwater floods reduced the salinity and density of the surface ocean in the North Atlantic, causing a reduction in the ocean's heat 'conveyor belt', known as the thermohaline circulation. This phenomenon is the large-scale ocean circulation that is driven by seawater density and wind. Wind-driven surface currents such as the Gulf Stream and its northerly extension, the North Atlantic Drift, travel from the equatorial Atlantic Ocean, distributing heat to high latitudes. The meltwater disrupted this flow.

Eventually, as the meltwater flux abated, the thermohaline circulation strengthened again and the climate recovered. The end of the Younger Dryas, about 11,500 years ago, was particularly abrupt. In Greenland, temperatures rose 10°C in as short a period as a decade.

SUMMARY

The purpose of this chapter was to show the effects that the geological carbon cycle (amongst other large cycles) can have on the environment. It illustrates that natural processes such as large-scale volcanism or methane release from seabed or soil hydrates are perfectly capable of altering the environment very radically. Different effects are also able to combine forces into positive feedback that propels environmental change into top gear or 'runaway' climate change.

Two of the examples I have chosen (for example, the Permian-Triassic and the PETM) appear to be close in form to the changes we might expect in the near future, because they occurred due to more greenhouse gases being in the atmosphere and due to global warming. Some, like the Younger Dryas, appear not to relate directly to global warming or greenhouse gases, but are examples of extremely rapid change that is nevertheless a consequence of warming. In the case of the Younger Dryas the discharge of large glacial lakes into the North Atlantic in a geologically extremely rapid event caused rapid change. This kind of change can of course occur on top of the background global warming, or be a result of it. Sudden changes like these can be seen as tipping points to be avoided at all costs, but are intrinsically hard to predict.

This brief summary of periods of rapid global warming illustrates that change did occur, including extinctions, changes in the hydrological cycle and in sea level. The subtle reader will also have noticed that geological boundaries, the boundaries that geologists use to define the bases of periods of geological time (for example, the Triassic), often coincide with climate changes and the mass extinctions that accompany them. This is no accident; often long before it was known that climate change occurred at these points, palaeontologists and stratigraphers had seen abrupt changes in the lineages of fossils. Often these are the first stages in the distinction of boundaries of geological periods because they afford an easy definition. Many of these long-established boundaries such as those of the Permian-Triassic, the Cretaceous-Tertiary and the Palaeocene-Eocene have now become the subject of intense research.

But beyond extinctions and changes in the hydrological cycle and in sea level, what real practical knowledge or advice can a deep-time climate-change scientist pass on to the people that really need to plan for modern climate change—for example, policymakers, regulators and governments?

There are a few problems in being able to do this. The first is the huge difference between the periods of time that these two types of specialists deal in. Policymakers and regulators look ahead weeks, months, and, if they have sufficient budget, years. Deep-time palaeoclimatologists deal in tens, hundreds, thousands and (mostly) millions of years. Could a palaeoclimatologist looking at the rock cores from the PETM provide a precise measure of sea level rise per degree of air temperature rise, or per ppm (part per million) of CO_2 rise that a coastal planner wondering about the life of local coastal defences would be able to use? Very unlikely.

A second reason for the lack of direct relevance of deep-time palaeoclimatology to planners and policymakers is the difference between the worlds that they are dealing with. The PETM world was very different to today. There was no Atlantic Ocean, the North Sea was enclosed to the south where the English Channel now divides England and France, and the PETM began at a time of already elevated temperature. Similarly, even though the end of the Younger Dryas about 11,500 years ago had temperature increases perhaps closest in rate of change to what we see today, the change began from a very different environmental starting point. Through the Younger Dryas, for example, sea ice covered the north of the North Sea and tundra covered much of England and Northern Europe. Also, of course, the reason for the rapid warming at the end of the Younger Dryas was different to what we see today—geomorphological and ocean current change rather than primarily atmospheric change.

Planners and policymakers are beginning to think about climate change through climate projections or scenarios which are derived from complex models that essentially extend weather forecasting, rather than feed in deep-time data. In the United Kingdom, for example, the UK Climate Projections 2009 (or UKCP09 for short) can be used. These projections provide relevant information about the implications of climate science and attempt to quantify the uncertainties of predictions. In general, UKCPO9 suggests warmer and wetter winters, hotter and drier summers, sea level rise, and more severe weather for the UK. Based on a 'medium emissions' pathway, which according to the UK Committee on Climate Change is the one that the world is currently most closely following, we could see average summer temperature increases in the South East region of England of 3.9°C by the 2080s. At the same time, we could see a 22% decrease in average summer rainfall in the South East and an increase of 16% in average winter rainfall in the North West by the 2080s, with increases in the amount of rain on the wettest days. I will look at these forecasts and their implications in more detail in a later chapter.

BIBLIOGRAPHY

Alvarez, W., 2009. The historical record in the Scaglia limestone at Gubbio: magnetic reversals and the Cretaceous-Tertiary mass extinction. Sedimentology 56, 137–148.

Alvarez, L.W., et al., 1980. Extraterrestrial cause for the Cretaceous–Tertiary extinction. Science 208, 1095–1108.

Benton, M.J., Newell, A.J., 2014. Impacts of global warming on Permo-Triassic terrestrial ecosystems. Gondwana Res. 25, 1308–1337.

Brand, U., et al., 2016. Methane hydrate: killer cause of Earth's greatest mass extinction. Palaeoworld 25, 496–507.

Cuffey, K.M., Clow, G.D., 1997. Temperature, accumulation, and ice sheet elevation in central Greenland through the last deglacial transition. J. Geophys. Res. 102, 26383–26396.

von der Heydt, A.S., 2016. Lessons on climate sensitivity from past climate changes. Curr. Clim. Change Rep. 2, 148–158.

Jin, Y.G., et al., 2000. Pattern of Marine Mass Extinction near the Permian-Triassic Boundary in South China. Science 289, 432–436.

Kender, S., et al., 2012. Enhanced precipitation and vegetation changes in the North-East Atlantic at the Palaeocene-Eocene boundary. Earth Planet. Sci. Lett. 353-354, 108–120.

Mathez, E., 2009. Climate Change: The Science of Global Warming and Our Energy Future. Columbia University Press. 368pp.

Pacala, S., Socolow, R., 2004. Stabilization wedges: solving the climate problem for the next 50 years with current technologies. Science 305, 968–972.

Richards, M.A., et al., 2015. Triggering of the largest Deccan eruptions by the Chicxulub impact. Geol. Soc. Am. Bull. 127, 1507–1520.

Skinner, L., 2008. Facing future climate change: is the past relevant? Phil. Trans. R. Soc. A 366, 4627–4645.

Sluijs, A., et al., 2007. The Palaeocene-Eocene thermal maximum super greenhouse: biotic and geochemical signatures, age models and mechanisms of global change. In: Williams, M., Haywood, A.M., Gregory, F.J., Schmidt, D.N. (Eds.), Deep time perspectives on Climate Change: Marrying the Signal from Computer Models and Biological Proxies. The Geological Society, London, pp. 323–349.

Steffen, W., et al., 2015. The trajectory of the Anthropocene: The Great Acceleration. Anthropocene Rev. 2, 81–98.

Twitchett, R.J., 2005. Climate change across the Permian/Triassic boundary. In: Williams, M., Haywood, A.M., Gregory, F.J., Schmidt, D.N. (Eds.), Deep time perspectives on Climate Change: Marrying the Signal from Computer Models and Biological Proxies. The Geological Society, London, pp. 191–201.

Zalasiewicz, J., Williams, M., 2016. Climate change through Earth's history. In: Letcher, T.M. (Ed.), Climate Change, second ed. Elsevier, pp. 3–17.

CHAPTER 3

Artificial Global Warming: The 'Fossil Economy'

Contents

In this chapter I examine the perturbations that humankind can inflict on the Earth's system, mainly through short-circuiting the long-term carbon cycle by converting geological carbon into atmospheric carbon dioxide. This involves huge (at least on the human scale) shifts in the way that human beings live, use the land, and generate electricity. For the carbon cycle, the most important first shift was the take-up of coal. In Britain, then home of the industrial revolution, coal had replaced wood as the main source of domestic heat in about 1700, but this was not an event that registered markedly on the atmosphere. It was the take up of coal industrially that was more important: from using natural falling water to generate power to spin cotton, to using the coal-powered steam engine. This was a big moment in human history, where human society stepped beyond natural limits and began to register on the composition of the 18th-century atmosphere. Shifts from one main energy source or 'prime mover' to another are known as energy transitions, and each of these has had an impact on the carbon cycle: wood to coal, coal to oil, oil to gas—and in the last few years in the United States conventional oil and gas to unconventional oil and gas.

The previous chapter illustrated the environmental chaos that natural perturbations can cause in the Earth's climate: for example, volcanic eruptions and methane release from hydrates. These natural perturbations can be pinpointed

Energy and Climate Change
https://doi.org/10.1016/B978-0-12-812021-7.00003-8

in rocks and ice cores and their progress can be plotted, admittedly only at timescales of hundreds, thousands, or millions of years. The beginning of the 'artificial', anthropogenic changes which modern climate scientists now describe started with the industrial revolution, or perhaps earlier, because this is the point where humankind began to use its fossil-fuel resources in a systematic and large-scale way. This was the birth of the 'fossil economy'.

THE 'FOSSIL ECONOMY': BRITAIN AND THE INDUSTRIAL REVOLUTION

The industrial revolution in Britain was a transition to organised, large-scale manufacturing processes between 1760 and 1840. It included changing from hand production to machines, new chemical and iron manufacturing, improved use of water power, the rise of the factory system—and, crucially for the purposes of this book, the uptake on a large scale of fossil fuels, mainly coal. Cotton manufacture was the earliest industry and its story is essentially one of transition from water-powered factories to those powered by steam.

Just before the industrial revolution, Britain had a rising population—the population doubled between the early 16th century and the mid-17th century—and many of these new people lived in towns far from the nearest wood fuel. Firewood increased hugely in price with the result that coal very soon became the cheapest fuel, especially in towns along the east coast of England, including London, which could get coal by sea from the North East (Fig. 3.1). In the early 17th century Britain's dependence on coal was already established. By 1700, coal had already overtaken wood as the leading provider of heat in homes.

Today most of Britain's coalfields are quiet, but coal has been mined in the central lowlands of Scotland, Northumberland and Durham, North and South Wales, Yorkshire, Lancashire, the East and West Midlands and Kent. The early exploited coalfields map onto the centres of the industrial revolution and their influence on wealth in the UK, and the influence of coalfields generally across the world, is clear because so many large modern industrial cities stand above coalfields.

Coal made its influence felt partly through manufacturing and steam power, and the development of cotton milling illustrates how water power was quickly replaced by steam power because of its adaptability and availability.

The first cotton mill to be powered by water was Marvel's Mill in Northampton in the English Midlands, which operated from 1742 until 1764. The entrepreneur Richard Arkwright developed water frame

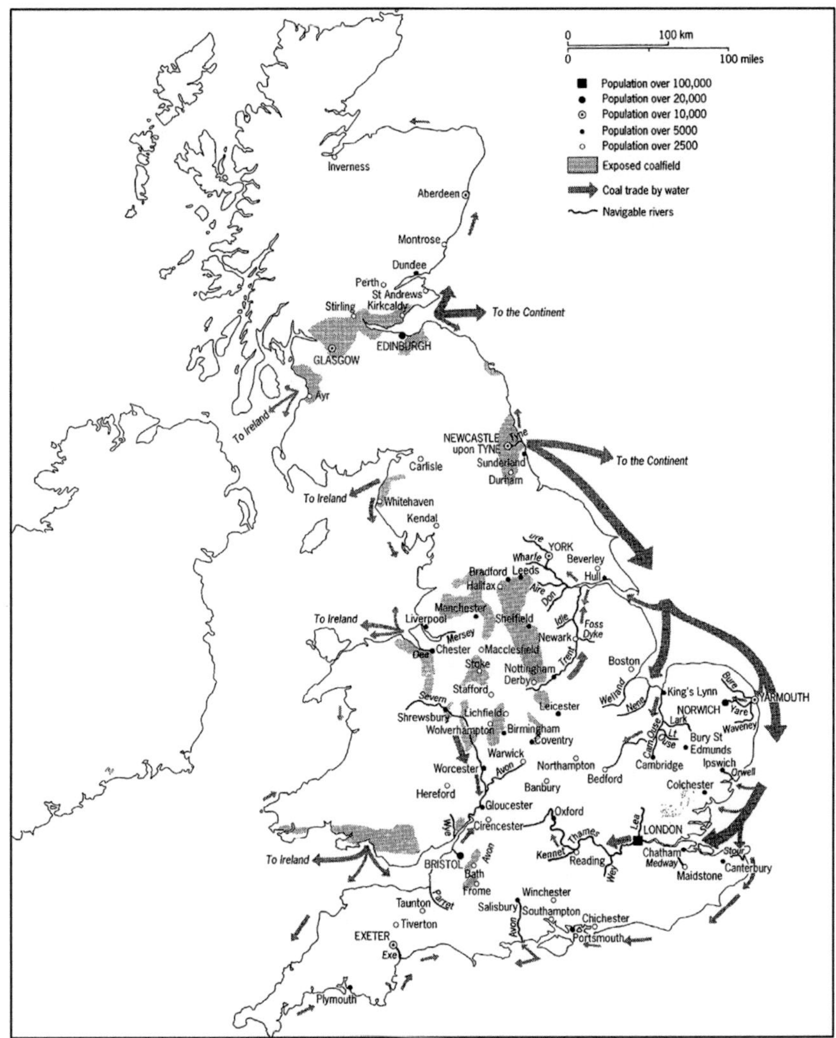

Fig. 3.1 Coalfields and trade routes in Britain in 1700. *(From Hatcher, J., 1993. The History of the British Coal Industry. Clarendon Press, Oxford.)*

spinning machinery in 1769 to make the water power more efficient, but it was not until Arkwright and his partners opened Cromford Mill in Derbyshire in 1771 that the first factory could be said to have been established (Fig. 3.2). It consisted of five storeys to which the water power was distributed, with large, well-lit internal spaces for workers. The form of Cromford Mill influenced cotton mills and other industrial architecture in Britain and elsewhere long after.

Fig. 3.2 Cromford Mill, the archetypal water-driven cotton mill, and perhaps the first modern factory. *(By chevin—Own work, Public Domain, https://commons.wikimedia.org/w/index.php?curid=6615061.)*

At the time of Cromford, only the kinetic energy of falling water provided enough power to drive large mills. But this 'prime mover' had the disadvantage of needing a constant flow of water. This meant that many mills were in upland areas away from centres of population. Mill owners needed to attract workers and some resorted to establishing workers' accommodation, effectively colonies. Steam power was naturally attractive to the same owners because it was not as dependent on the vagaries of the weather and did not have to be established on rivers with a convenient 'fall'. Steam-powered mills could be established where there were coal and workers.

Steam engines had been used before to pump supplementary water to the wheels of cotton mills. But in 1781, James Watt registered a patent for the first rotative steam engine that could directly supply power. Even then it was not until 1785 that a steam engine was successfully used to drive a cotton mill, at Robinson's Mill near Nottingham.

After this innovation, steam engines allowed mills to be established in urban areas of plentiful labour and trade connections. Amongst other innovations, this transformed northern cities like Manchester, which had been a centre of hand spinning and weaving. By 1800 Manchester had 42 mills, and was the heart of the world's cotton manufacturing trade.

The rotative steam engine essentially created a demand for coal apart from domestic heating, but also allowed mines to go deeper because steam engines could drain mines of water. Steam engines also drove large fans to clear gas in mines. Between 1769 and 1800, Britain's annual coal production doubled.

Coal use on an industrial scale spread rapidly, mainly driven by manufacturing. In Germany mining on a large scale began in the 1750s in the Ruhr and Wurm valleys; similar iron and steel development occurred in Wallonia in Belgium. Further east in Poland, Silesia developed coal mining at a rapid pace. In the northeast of the United States, coal had become the main domestic fuel of the cities by 1850. This is how the 'fossil economy' started.

The phrase 'fossil economy' was used by the historian Andreas Malm in the book *Fossil Capital: the rise of steam power and the roots of global warming*. Malm relates the uptake of fossil fuels and global warming but also capitalism—the sorts of money flows that can afford massive investments relating to coal mining, steam production, and cotton manufacture—to cotton, the industrial revolution and the concentration of population in cities. Essentially, he sees coal and steam as allowing the natural supply of non-fossil energy to be bypassed. It was a point when human society began to live 'beyond its means', when the natural bounds of nature were overstepped:

> *The original purpose of coal—heat for the populace—opened a hallway to population concentrations, which subsequently lured manufacturers away from water as a source of mechanical energy in a historical cunning of sorts. Coal in stoves contributed to the pattern of centralised settlements; water mills came into contradiction with this pattern; the conversion to steam resolved it by bringing capital and labour together.*
>
> ***(Malm, 2016)***

In contrast to our modern views, in the 18th and 19th centuries steam and coal began to be seen as a moral force for good:

> *The steam engine consequently as the slave of man, is the machine which is most highly valued and which under skilful direction has accomplished more than any other machine for the promotion of the comfort, convenience and wellbeing of mankind…'*
>
> ***(The Industry of Nations, 1855, Part II)***

Poetry and paintings celebrated the new world of steam. Joseph Turner's painting of 1844, 'Rain, Steam and Speed', glamourized the new technology of the railways in Britain, showing a new rail bridge across the Thames River, conveying speed and steam (Fig. 3.3).

Fig. 3.3 'Rain, Steam and Speed—The Great Western Railway', an oil painting celebrating the world of steam by the 19th century British painter Joseph Mallord William Turner.

EARLY ATMOSPHERIC EFFECTS OF THE LARGE-SCALE TAKE-UP OF COAL

This intense economic and social development clearly had effects on the environment including local pollution. But modern studies also show the effect of the new industry on the atmosphere. Direct evidence that the large-scale take-up of coal affected the atmosphere comes from the isotopic composition of the carbon in the CO_2 of the atmosphere through time (Fig. 3.4).

The graph shows that in the last 200 years the $\delta^{13}C$ of CO_2 in the atmosphere has decreased by 1.5‰. Ice core data shows the trend across a greater time period (from 1700 AD) with a particular increase in the steepness of the curve at about 1800, not long after the start of the industrial revolution.

The reason for this change is partly a matter of dilution of atmospheric CO_2 with older carbon in CO_2 released from fossil fuels. The carbon in CO_2 that is released from burning coal, gas, and oil has much lower $\delta^{13}C$ than atmospheric CO_2. For example, the carbon in 300-million-year-old Carboniferous coal has a $\delta^{13}C$ of between −23‰ and −24‰ because this was the $\delta^{13}C$ in the plant tissues that formed the coal. This is much lower than the $\delta^{13}C$ of the CO_2 in the modern atmosphere (about −8‰) and so when the low $\delta^{13}C$ CO_2 of the burned Carboniferous coal enters the atmosphere, the $\delta^{13}C$ in the mix of modern atmospheric CO_2 decreases.

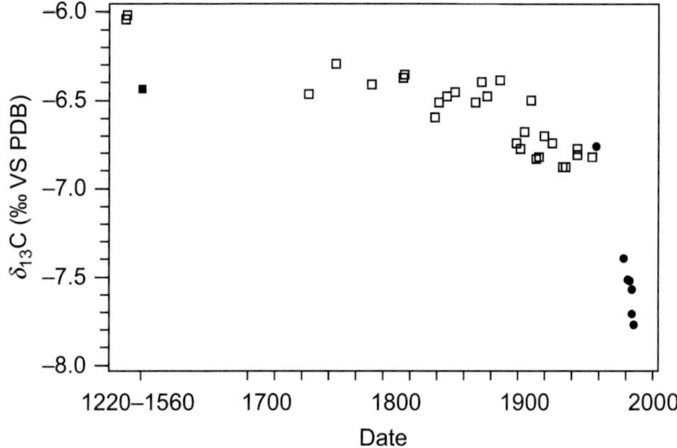

Fig. 3.4 $\delta^{13}C$ records of the carbon in atmospheric CO_2 from Mauna Loa Observatory and ice core from South Pole ice. The trend is clearly decreasing. Circles from Mauna Loa Observatory and squares from South Pole ice. *(From Ghosh, P., Brand, W.A., 2003. Stable isotope ratio mass spectrometry in global climate change research. Int. J. Mass Spectr. 228, 1–33.)*

More circumstantial evidence of the connection between fossil fuels and CO_2 in the atmosphere comes by simple comparison of historical data on fossil-fuel burning and CO_2 concentration. This was shown by David MacKay in his excellent book *Sustainable Energy without the Hot Air.* Mackay simply drew graphs of coal and oil use on the same horizontal axis as for CO_2 concentration (Fig. 3.5).

The similarity of the curves is striking. The upper curve—the CO_2 'hockey stick'—has a counterpart in the lower curve. Also striking is the relationship with advances such as the development of the steam engine.

The same approach of looking at historical emissions can be used to ascertain cumulative emissions for different countries or parts of the world. Britain and Germany are amongst the countries that top the historical cumulative emissions chart (Fig. 3.6), reflecting their early take-up of coal in the industrial revolution. Expressed as an average emission rate over the period 1880 to 2004, the United States for example had a rate of almost 10 tons of CO_2 per person per year, though not all of this relates to the use of coal, as we will see in the next section.

THE GROWTH OF OIL

If the growth of coal was a British story, then the growth of oil is an American story—of rapid industrialisation and corporate and geopolitical power. The

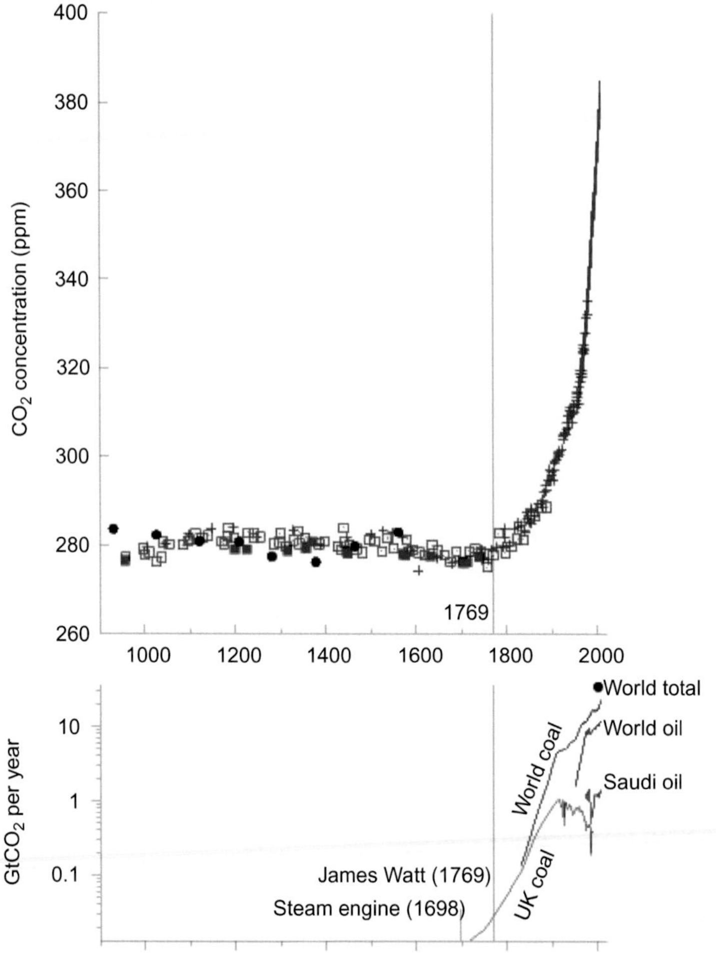

Fig. 3.5 Atmospheric CO_2 concentration and historical fossil-fuel usage. The vertical axis of the lower graph uses the gigatonnes of CO_2 per year ($GtCO_2$ per year) unit rather than the number of barrels or tonnes of oil and coal used. *(From MacKay, D., 2009. Sustainable Energy Without the Hot Air. UIT.)*

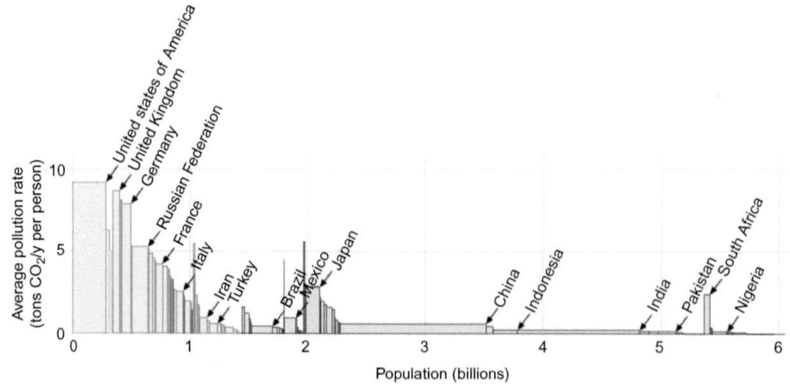

Fig. 3.6 Historical emissions of nations between 1880 and 2004 (average per person per year). The width of the rectangles representing countries shows the sizes of their populations. *(From MacKay, D., 2009. Sustainable Energy Without the Hot Air. UIT.)*

'oil age' has lasted around 150 years, starting around 1860, and essentially involved the replacement of coal by oil as the primary energy provider, though coal still dominates the generation of grid electricity. Presently oil in its various forms provides 90% of fuel for all vehicles. It also provides 40% of primary energy across the world, but still is responsible for only 2% of electricity generation. Its main value is as a portable, energy-dense fuel.

Long before the industrial revolution, naturally occurring bitumen was used in construction, waterproofing for the hulls of boats, and in medicine, mainly in the Middle East. In the same area, crude oil was also distilled on a limited scale for lamp oil. The natural occurrence of oil in the northeast of the United States in Pennsylvania also led to a small medicinal oil business in the early 19th century. But the lack of whale oil for lamps began to increase the need for an alternative source, and simple refining showed that mineral oil could provide it.

The first concerted successful effort to drill for oil (as opposed to getting oil accidently after drilling a water well) was on 27 August 1859 at Oil Creek near Titusville in Pennsylvania, for the lamp oil market. This new supply was rapidly taken up and the Titusville and other wells produced 2000 barrels in 1859, but 4.8 million barrels by 1869.

This mineral oil was primarily refined for lamp oil, and other fractions of the oil were often discarded. The next phase of oil was the realisation that these fractions could be used in other ways, primarily in the internal combustion engine. So as oil is an American story, it is also a story of the automobile and petroleum refining, including names such as John D. Rockefeller and Henry Ford. The early oil business was dominated by Rockefeller's Standard Oil Company, which by 1878 controlled more than 90% of the total American refining capacity, and this concentration of monopolistic power was part of the reason for the US government disbanding Standard into several smaller companies in 1911.

Until 1900, the oil business had been concentrated in the Northeast and Midwest, but the discovery of large amounts of oil at Spindletop in the Texas Gulf Coast changed much, in that oil suddenly came from a different area, and fostered the development of companies that were not related to Standard Oil and which could provide fuel oil, for example for ships in the Gulf Coast. At about the same time, the automobile began to be mass produced, beginning with the Model T Ford in 1908. Within a few years, the US car industry was producing more than 500,000 cars a year. The introduction of thermal cracking and rotary drilling meant an enormous growth in the production of oil between 1900 and 1920 to feed the automotive and other fuel industries.

In early foreshadowing of the present 'peak oil' debate, the US Geological Survey in 1908 predicted total exhaustion of US oil by 1927. This kind of recurring concern over supply was related partly to the boom and bust engendered by the cycle of field discovery and depletion. But by 1930, oil's share of total US energy was 23.8%, mainly due to gasoline consumption.

Around the 1930s the oil business began to globalise. US companies obtained drilling licences in Iraq, Bahrain and Saudi Arabia, and geopolitics began to surface, for example the strategic need for oil which had implications for World War II in terms of territorial land grabs of prospective areas.

As well as the coincidence of the success of early drilling, a strong gasoline market, and vigorous business environment, a series of technological advances helped to boost production and better understanding of the habitat of oil in the subsurface. In early drilling, oil was thought of as moving around in underground rivers and caverns.

Some of these technologies included the first water-based drilling mud and steam-driven rotary drills around the turn of the century, and in 1909 the first roller-cone drill bit. To collect better data from the subsurface, electrical logging began to be used in new wells around 1927 for probing the geological properties of rocks, about the same time that reflection seismology began to probe the structure of the subsurface.

In the early 1940s, horizontal well drilling was developed in the Soviet Union and this was later followed in Stephens County, Oklahoma and Archer County, Texas in 1949 by hydraulic fracturing. These techniques combined became key to the huge increase in shale gas production in the early 2000s in Texas.

In the early 1980s, 3D seismic processing began allowing a much more sophisticated way to image subsurface reservoirs and to plan well drilling. In 1996 Qatar opened the world's first major liquid natural gas (LNG) exporting facility.

Today the top three oil-producing countries are Saudi Arabia, Russia, and the United States. About 80% of the world's resources are located in the Middle East, with around 60% of those coming from the 'Arab 5': Saudi Arabia, UAE, Iraq, Qatar and Kuwait. However, high oil prices can make heavy oil and oil sands, for example in Venezuela and Canada, prospective also.

THE GROWTH OF GAS

The growth of natural gas as a fuel in many ways is mirrored by that of oil, partly because oil is produced alongside gas in many cases. However, gas was held back more than oil in that its transport in useful quantities, in the early years at least, was governed by pipelines. The early uptake of natural gas in the United States therefore tends to relate to technological and infrastructure improvements in pipelines, just as its global expansion today relates to developments in liquefied natural gas technology (LNG) and infrastructure.

In the 19th century, manufactured coal gas dominated gas for domestic use in cities in Europe and North America. In the 1860s natural gas began to be used for local heating, but it was not until the 1870s that gas began to be used on a city scale, for example in Rochester, New York, where a 25-km pipeline was built. Waste gas from oil wells in the Titusville area began to be used in that town around 1872.

In an interesting echo of the present substitution of gas for coal in power stations to reduce emissions, the city of Pittsburgh looked in the 1880s to increase the use of gas in steel and iron manufacture to reduce air pollution. Pressurised pipelines moved gas from fields such as the Murrysville gas field to the city, so that by 1890 Pittsburgh was the centre of the iron, steel and natural gas industries.

Natural gas was discovered in the Texas panhandle in 1918 such that it shortly became the richest natural gas area in the United States, but again the lack of transportability caused a hurdle until in the 1920s pipeline construction and gas compressor technology made long-distance pipelines a possibility. This meant that gas could be supplied to the Northeast, a situation that was not reversed until the 2000s, when shale gas from Pennsylvania began to be supplied in the opposite direction.

Shale gas in the United States has been the surprise development in petroleum in the present century. Shale gas began to pick up in the early 2000s to become about a quarter of US gas domestic production. Shale is forecast to produce around half of US domestic gas by 2040 (Fig. 3.7).

Shale gas had its roots in the 1973 and 1979 oil shocks that led the United States to address energy shortages and the high price of oil by investing in research and demonstration of other sources of energy, including shale gas. Amongst the most important of the government research initiatives was the US Department of Energy's (DOE) Eastern Gas Shale Project, which looked at the possibilities of shale gas in Illinois and Michigan. The Gas Research Institute (GRI) was established in 1977 and in the 1980s and early 1990s it started research programs on transmission, distribution, and markets of gas.

But a very vigorous private sector had a strong role also. In the 1980s and 1990s around 10,000 wells were drilled around Fort Worth in north Texas. A combination of horizontal drilling which followed the rock layers and low-cost hydraulic fracturing made extraction economic.

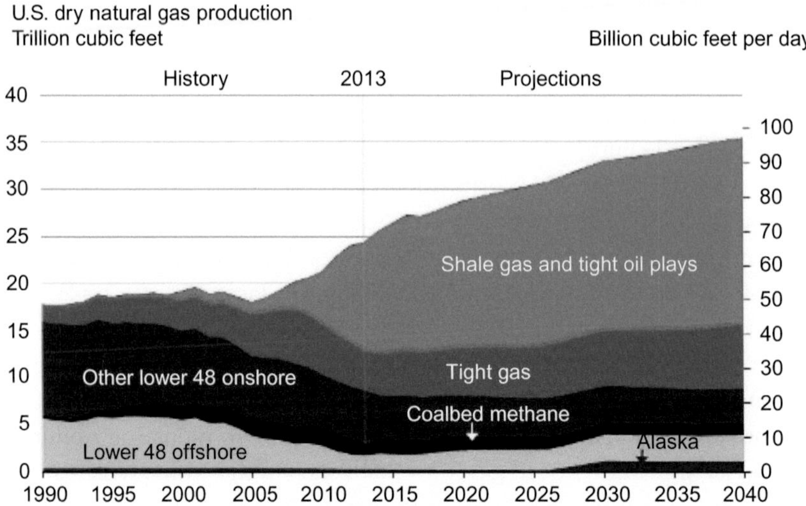

Fig. 3.7 US natural gas production, 1990–2040 (trillion cubic feet per year). *(From the EIA's Annual Energy Outlook 2015 Reference case.)*

Much of the early progress in shale gas was made by small companies like Mitchell Energy, but in the early 2000s larger oil and gas companies entered the business, mainly by buying up the small companies. Devon Energy bought up Mitchell Energy and Exxon Mobil bought XTO. These new larger companies were responsible for much of the development of Texas shale. Between 1997 and 2009, more than 13,500 gas wells were drilled—mostly of the horizontal variety—in the Barnett shale.

The US industry has been boosted by shale gas. The EIA Annual Energy Outlook of 2014 predicts a jump in manufacturing powered by cheap fuel. The main industries affected will be energy-intensive bulk chemicals and primary metals, both of which provide products used by the mining and other downstream industries, such as fabricated metals and machinery. The bulk chemicals industry is also a major user of natural gas and, increasingly, hydrocarbon gas liquids, which are often produced with gas from shale. One of these is ethane, which is used to make ethylene, which in turn has many industrial products including PVC, polystyrene, latex, detergent and vinyl.

Canada is the next most developed shale gas area after the United States, with about 3 billion cubic feet of gas being produced per day in 2013, mainly from the Muskwa shale in the Horn River area in northeast British

Columbia, the Montney shale in British Columbia and Alberta, and the Duvernay shale in Alberta. Some of the most productive of the American shale layers also extend under the international border into Canada, including the Antrim, Utica, and Marcellus shales.

Shale gas holds some fascination for Europe, partly related to the need for increased energy security and stringent, legally binding targets for greenhouse gas emissions. Europe gets 24% of its gas from Russia, and half of that—80 billion cubic meters (bcm) a year—passes through Ukraine. A disagreement between Russia and Ukraine led to the pipelines shutting down for two weeks in January 2009, causing worries across Europe. Events in Crimea in 2014 increased those fears. EU countries have 36 bcm of stored gas and could store 75 bcm—but even that could run out and ultimately its top-ups would mostly come from Russia. For Estonia, Latvia and Lithuania the situation is even worse, because these countries get all their gas from Russia (Fig. 3.8). Bulgaria gets almost all its gas from a Russian pipeline that crosses Ukraine, and it has limited storage.

In Britain, the energy security story is different. Britain has more diverse energy sources than most of Europe. The UK used to obtain electricity from three big sources—coal, gas and nuclear—but recently coal has declined and renewables and gas have increased. Most gas burnt in British power stations is imported from Qatar (as LNG) and Norway. Security of supply is still a concern because, as British North Sea oil and gas declines, dependency on imports will increase. Conventional oil and gas production from the North Sea passed its peak just before 2000.

Britain also has a very ambitious target of greenhouse gas reduction of 80% by 2050, meaning that much more electricity (for example) has to

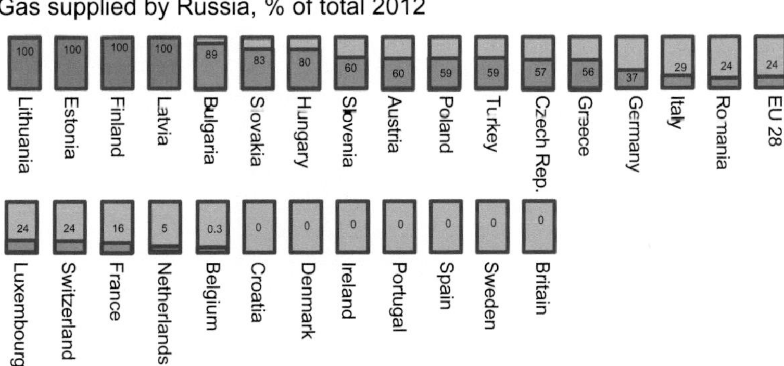

Fig. 3.8 Dependence on Russian gas in Europe. *(Source: The Economist.)*

be provided by low-carbon sources. Nuclear and wind are perhaps the most important of these and, while wind is developing, it is expensive. More expensive still is nuclear. Britain has a long history of nuclear and was one of the first countries to develop commercial nuclear. However, the cost of building and decommissioning power stations—and the public's dislike of them—has made their development more and more difficult.

EFFECTS OF THE TAKE-UP OF OIL, AND THEN GAS, ON GLOBAL EMISSIONS

In the same way that the emissions of CO_2 are related to industrial and large-scale coal use, the dominance of oil and gas can also be connected to marked changes in the atmosphere and indirectly to many other environmental changes. The mantle of biggest primary energy producer was taken from coal by oil and gas in the 1950s (Fig. 3.9). Increasingly this is seen as the start of the so-called 'Great Acceleration' which has great interest for the concept of the Anthropocene, which I will look at in a later chapter.

The Great Acceleration is seen as a transformation of the human relationship with the natural world, partly because so many activities started or increased rapidly at about 1950. Some of these activities were convincingly illustrated in a series of 12 graphs in 2004 by Will Steffen and colleagues. The updated graphs to 2010 are shown in Fig. 3.10.

A number of the changes shown are directly related to the uptake of the energy in oil and gas, for example transportation and international tourism.

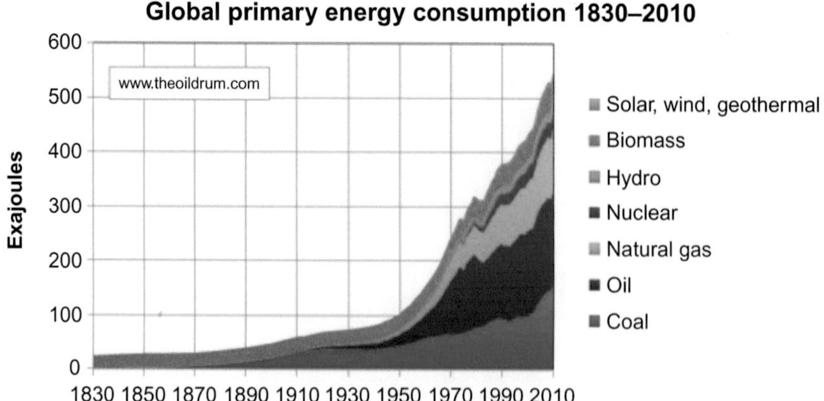

Fig. 3.9 Global primary energy consumption. *(From http://www.wcoes.org/2012/02/human-energy-consumption-moves-beyond.html.)*

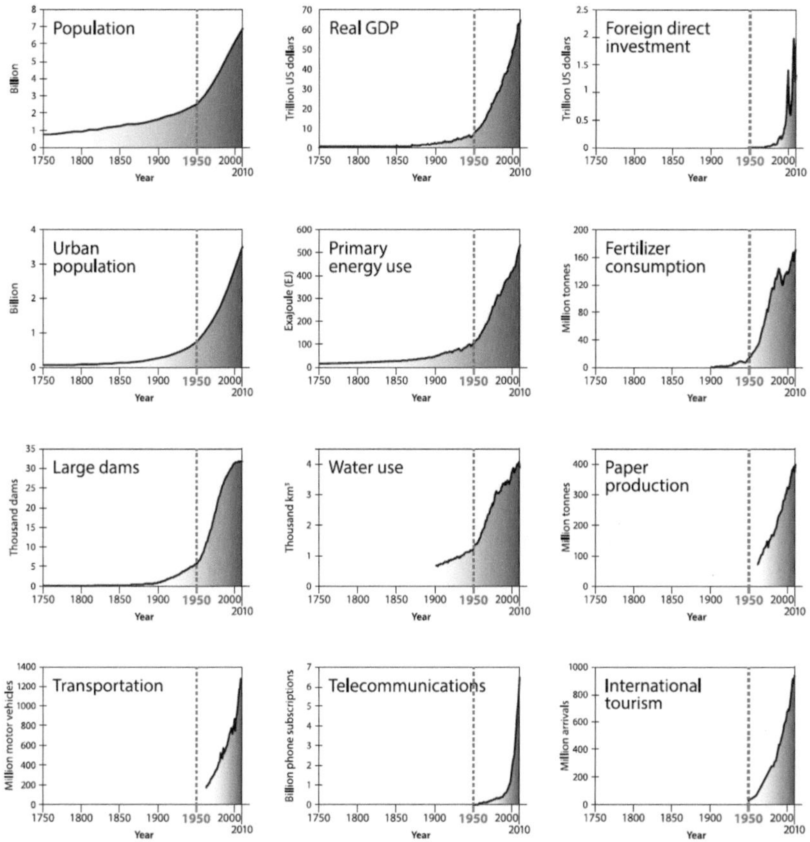

Fig. 3.10 Trends from 1750 to 2010 in globally aggregated indicators for socioeconomic development. *(From Steffen, W. et al., 2015. The Trajectory of the Anthropocene: The Great Acceleration. Anthropocene Rev. 2, 81–98.)*

Others are less directly related but use energy ultimately derived from oil and gas, for example large dam construction and paper production.

As for direct effect on the atmosphere of the increase in oil and gas usage at the beginning of the Great Acceleration, a sudden change in gradient is more difficult to pinpoint, with inflections possibly occurring around 1950 or, as in Fig. 3.11, in 1965. The graph of $\delta^{13}C$ also shows an inflection.

A more recent emissions change that has been mooted has been that associated with the uptake of shale gas in US power stations. According to the IEA's 2011 World Energy Outlook, it is fairly well established that shale gas substituted massively for coal in US power stations following the shale gas boom in the early 2000s, such that for the years 2006–2011, CO_2 from

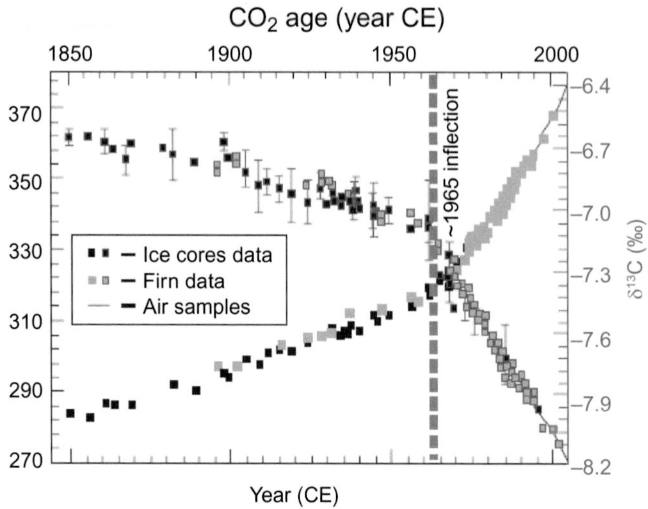

Fig. 3.11 CO_2 concentration and $\delta^{13}C$ from atmospheric CO_2 from the Law Dome ice core, firn data, and air samples. Firn is a compact form of snow. *(From Waters, C. N. et al., 2016. The Anthropocene is Functionally and Stratigraphically Distinct from the Holocene. Science 351, 137–147.)*

fossil-fuel consumption declined by 430 million tons (or 7.7%) in the United States, although U.S. energy-related CO_2 emissions increased from 5267 million tons in 2012 to 5396 million tons in 2013.

This reduction in CO_2 emissions has been championed by the shale gas industry in particular, as one of the advantages of 'home-grown' gas, even though coal that was not burned in US power stations reputedly found its way into other power stations, for example in Europe, so emissions could be said to have been only shifted geographically, rather than avoided.

Another interesting take on the question is whether greenhouse gas emissions (as opposed to just CO_2) have been reduced by the advent of shale gas in the US. In 2011 a team led by Robert Howarth of Cornell University studied the methane emissions from the surface installations of several shale gas wells, finding high levels of methane release into the atmosphere as a result of two routine processes: flowback and drillout. Flowback is the liquid that returns to the well and then to the surface following hydraulic fracturing. Typically, it carries methane. The early practice of leaving the flowback water in open tanks at the surface (ready for disposal or reuse) allows methane to escape directly into the atmosphere. After drill out (where plugs in the well that partitioned parts of the well for hydraulic fracturing are removed), methane also escapes from the well to the atmosphere. Howarth published data that showed a very high level of methane leakage of between 3% and 8%

of a well's production over its lifetime. Given that methane is a potent greenhouse gas, he considered that such levels of emissions cast doubt on the claims that shale gas could be a low-carbon fuel, for example in comparison with coal. Howarth's paper has been contested and recent work has suggested that the high figures for emissions quoted in the paper were unrepresentative of the industry as a whole; for example, a large study by Phillip Allen and colleagues in 2014 suggested that the leakage rate is about half of 1% of gas production. The difficulty of establishing with certainty what the net balance of emissions from different sources means for the atmosphere shows how large-scale transitions in energy need to be carefully monitored and understood.

OIL, GAS, AND COAL RESOURCE AND RESERVE, AND 'UNBURNABLE' CARBON

Oil and gas subsurface terminology includes two important terms: resource and reserve. The two are similar in that they describe amounts of oil and gas (or any other useful mineral) under the ground, but the difference between them is very important. The oil and gas *resource* for a country or a region is the total amount that in geological terms constitutes material of potential value. The idea of *reserve* is very different. Reserve is the proportion of the resource that is economical to produce, within environmental and social limits. Another concept, the *recovery factor*, connects the two in that it is the percentage of the resources that can become reserves.

Fig. 3.12 illustrates the difference between the two concepts as a flow chart. In order to realise a resource as a reserve, a number of requirements have to be satisfied, such as compliance with regulations.

The one requirement that I have left off the diagram is the concept of 'unburnable' carbon. I will look into the details of the United Nations Framework Convention on Climate Change (UNFCCC) Paris Agreement in a later chapter, but it has a special significance for the resource and reserve of world fossil fuels; this is because to abide by the commitments of the Paris Agreement, some fossil fuels cannot be burned. They have to be left in the ground, unmined and undrilled.

Research by the Potsdam Institute in Germany showed that to reduce the chance of exceeding 2°C warming to 20%, the global carbon budget, or the amount of fossil fuel we have left to burn between 2000 and 2050, is 886 $GtCO_2$ (gigatonnes of CO_2). If the emissions of the first decade of this century are subtracted from this, because they have already been emitted, this leaves a budget of 565 $GtCO_2$ for the remaining 40 years to 2050.

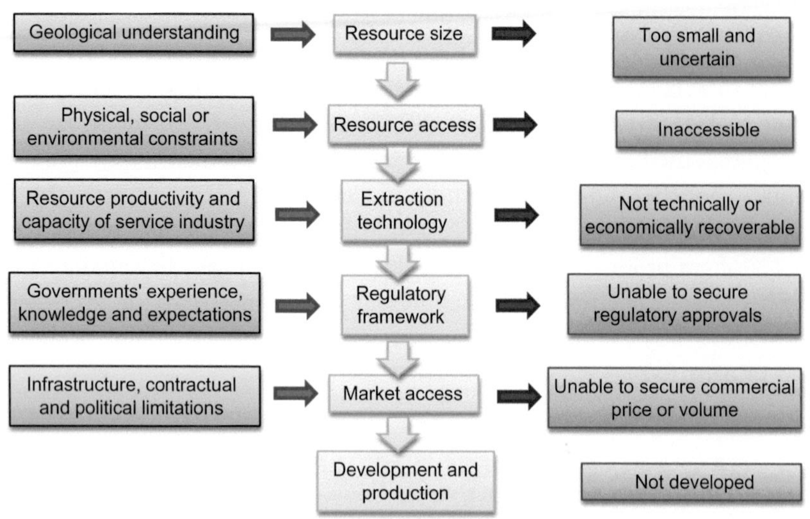

Fig. 3.12 The stages by which a resource is turned into a reserve. *(From Stephenson, M.H., 2015b. Shale gas and fracking: the science behind the controversy. Elsevier, Amsterdam, 153pp.)*

The fossil fuel proven reserves of the largest 100 stock exchange–listed coal, oil, and gas companies represent emissions of 745 $GtCO_2$. Obviously, this exceeds the budget of 565 $GtCO_2$, which means that should listed companies realise these reserves, emissions will go beyond 2°C of global warming. It also means that companies cannot burn all their reserves. Technically, this could mean that a large proportion of these listed companies' fossil-fuel assets are not reserves at all, because they cannot be produced within regulatory rules. Including other government-owned companies that hold reserves (for example, Arabian Gulf national oil companies) makes the situation even worse, indicating even more unburnable carbon.

Apart from the problem that this creates for climate change, it also means that investors and shareholders in companies are exposed to the risk of unburnable carbon and that companies are potentially overvalued. This has far-reaching implications for pension funds, many of which invest in fossil-fuel companies. Many fear that unburnable carbon will make countries or companies with the biggest reserves look at their assets in a rather different way. If these countries are not able to realise their value in the long term as carbon targets become more serious and binding, then the feeling may be that they have to drill and sell those assets now. Hydrocarbon-rich countries like the Arabian Gulf countries, Venezuela, Russia, Nigeria, and Angola, which a few years ago confidently looked ahead to many years of production and sales, now see a more problematic future.

THE TRANSITION OF ENERGY SYSTEMS

This chapter has attempted to survey how energy systems emerge, become dominant, and then fade. Transitioning from the current 'oil economy' to a lower carbon renewables economy is one of the biggest challenges for the modern world, alongside challenges such as communicable diseases and urbanisation. The chapter has shown that this kind of transitioning does occur and that there are various reasons for transitions, not all simply economic, such as the switch from coal to natural gas in the Pittsburgh steelworks of 1870. What gives the needed transition such extra imperative is the fact that to offset the intrinsic momentum of climate change, modern society needs to act before many of the most serious environmental changes become obvious. Apart from being a difficult sell to a partly-sceptical public, this also needs careful planning and a good understanding of past transitions.

Fig. 3.13 is a useful way to look at energy transitions. It shows the percentage global supply of different fuels and illustrates how wood or biomass were supplanted by coal, and coal by crude oil. Natural gas is perhaps the next feature in the energy landscape and is advocated by some as a cleaner fuel than both oil and coal, as well as a 'bridge fuel' to renewables. Time will tell whether gas gains the dominance of oil, but in many countries gas is already the mainstay of baseload electricity and heating for homes.

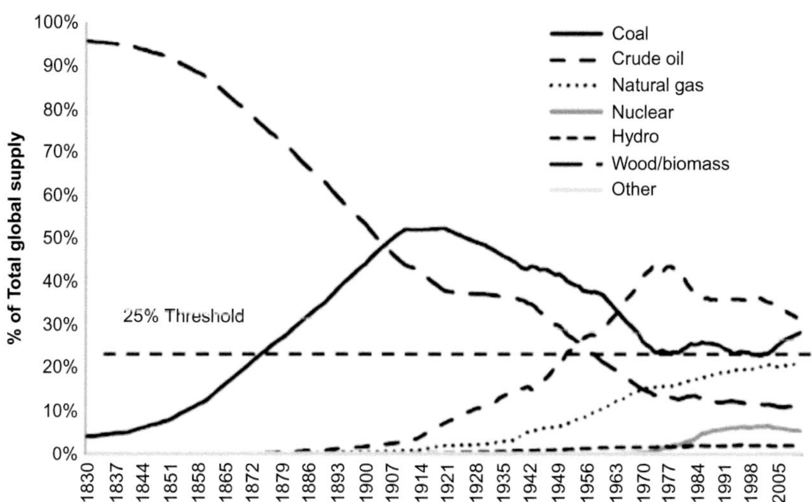

Fig. 3.13 Global energy supply by fuel source as a percent of the total, 1830–2010. Note "Wood/Biomass" includes biofuels, and "Other" includes renewable sources of energy such as wind, solar, and geothermal. *(From Sovacool, B.K., 2016. How long will it take? Conceptualizing the temporal dynamics of energy transitions. Energy Res. Social Sci. 13, 202–215.)*

An article by the energy policy expert Benjamin Sovacool looks in detail at the ways that these and other transitions occur and how long they take. A transition could be seen to have occurred if the incoming fuel reaches, for example, 20% or 50% of the market, and it can be made up of a series of smaller changes. The ascent of oil, for example, could be seen as: the switch from animal power to internal combustion engines; the conversion of steam engines to diesel on ships and locomotives; the shift from candles and kerosene to oil-based lamps; the adaptation of coal boilers to oil boilers for electric power; and the exchange of wood and coal stoves for gas furnaces in homes.

Sovacool mentions the notion of innovations moving out from a core (the innovative first adopter), to the rim (early adopter), to the periphery (late adopters). In the case of the replacement of traditional wood to coal, the core as discussed early in this chapter was Britain, while the rim countries (Table 3.1) were Germany, France and the Netherlands. The speed at which these energy transitions took place varied from 160 years in the case of Britain to 102 years in Germany. According to Sovacool, amongst the stages that are experienced are: a period of extended experimentation with small-scale technology and a diversity of design, followed by scale up of particular units of technologies as designs improve and economies of scale emerge, and by scaling up at the industry level. As industry structure becomes standardised and core markets become saturated, further industry growth is driven by globalisation, the diffusion of a successful design from the innovation core to rim and periphery markets.

These stages can be seen in the US shale gas boom, with experimentation in the Barnett shale occurring in the 1990s, and in the early 2000s improvements in the techniques of hydraulic fracturing. Perhaps the final

Table 3.1 Speed of the wood-to-coal energy transition in Europe

Wood to coal		Diffusion midpoint	Diffusion duration (yrs)
Core	England	1736	160
Rim	Germany	1857	102
	France	1870	107
	Netherlands	1873	105
Periphery	Spain	1919	111
	Sweden	1922	96
	Italy	1919	98
	Portugal	1949	135

(After Sovacool, B.K., 2016. How long will it take? Conceptualizing the temporal dynamics of energy transitions. Energy Res. Social Sci. 13, 202–215.)

globalisation stage is just beginning with shale gas starting to develop in China and Argentina, which are 'rim' countries.

Sovacool's main conclusion appears to be that transitions vary in duration. In the United States crude oil took half a century from its exploratory stages in the 1860s to capturing 10% of the national market in the 1910s, then 30 years more to reach 25%. Natural gas took 70 years to rise from 1% to 20% in the United States. Coal needed 103 years to account for just 5% of total energy consumed in the United States and an additional 26 years to reach 25%. Though other transitions appear long and drawn out, some are quicker, for example the take-up of nuclear power in France, which took 11 years to get to a 25% share of the market.

A final consideration is the fact that some transitions may be primarily driven by scarcity and the attendant effects of price and inertia. Since the early worries of the US Geological Survey about exhaustion of US oil by 1927, 'peak oil' has been predicted on numerous occasions, but then each time extended further into the future due to technological advances. Recent curves suggest coal phase-out around 2050, oil (phase out) around 2020, and gas around 2040. So all fossil fuels should peak roughly within the next half century (Fig. 3.14).

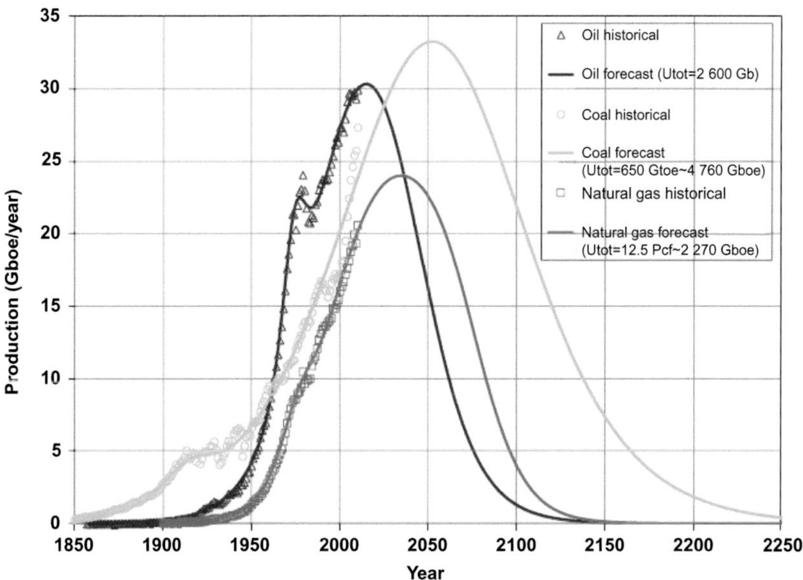

Fig. 3.14 Comparison of fossil fuel forecasts. Fossil fuels should peak roughly within the next half century. *(From Maggio, G., Cacciola, G., 2012. When will oil, natural gas, and coal peak? Fuel 98, 111–123.)*

SUMMARY

Human energy transitions have a strong effect on the global carbon cycle, alongside other human activities such as changes in land use. The results of this are measurable changes in the atmosphere detectable both in ancient ice cores and by direct measurement of the modern atmosphere. Perhaps the largest of these changes in the CO_2 concentration in the atmosphere in the late 18th century can be ascribed to the industrial revolution and the large-scale take-up of coal as a fuel. This 'inflection point' on the CO_2 curve marks an important change in humankind's relationship to natural resources, according to the historian Andreas Malm, who believes that at this point energy resource switched from being provided at the surface of the Earth to being provided from below ground. The coal used was outside of the short-term carbon cycle, and allowed the industrialists to tap into an ancient store of carbon, not previously used systematically or on a large scale. Later transitions, for example from coal to oil, and perhaps from oil to gas, have continued to cement our relationship with fossil fuels such that perhaps an even more profound 'inflection point' has been passed, known as the Great Acceleration, a point where many industrial activities started or increased rapidly.

The big unknown at the moment is the way that the developing world will industrialise and find its energy. This is examined in the next chapter.

BIBLIOGRAPHY

Carbon Tracker Initiative, 2017. Unburnable Carbon—are the world's financial markets carrying a carbon bubble? 33pp.

Energy Information Administration, 2014. Annual Energy Outlook 2014. early release.

Energy Information Administration, 2015. Annual Energy Outlook. 154pp.

Fouquet, R., 2015. Handbook on Energy and Climate Change. Edward Elgar Publishing Ltd. 752pp.

Ghosh, P., Brand, W.A., 2003. Stable isotope ratio mass spectrometry in global climate change research. Int. J. Mass Spectr. 228, 1–33.

Giebelhaus, A.W., 2004. History of the Oil Industry. In: Encyclopedia of Energy. 4, Elsevier Inc., pp. 649–660

Hatcher, J., 1993. The History of the British Coal Industry. Clarendon Press, Oxford.

International Energy Agency, 2011. World Energy Outlook, Paris, 696pp.

Karl, T.L., 2004. Oil-Led Development: Social, Political, and Economic Consequences. In: Encyclopedia of Energy. 4, Elsevier Inc, pp. 661–672.

MacKay, D., 2009. Sustainable Energy Without the Hot Air. UIT.

Maggio, G., Cacciola, G., 2012. When will oil, natural gas, and coal peak? Fuel 98, 111–123.

Malm, A., 2016. Fossil Capital: The Rise of Steam Power and the Roots of Global Warming. Verso, 496pp.

Salameh, M.G., 2004. Oil Crises, Historical Perspective. In: Encyclopedia of Energy. 4, Elsevier Inc., pp. 633–648
Sovacool, B.K., 2016. How long will it take? Conceptualizing the temporal dynamics of energy transitions. Energy Res. Social Sci. 13, 202–215.
Steffen, W., et al., 2004. Global Change and the Earth System: A Planet Under Pressure. The IGBP Book Series. Springer-Verlag, Berlin, Heidelberg, New York. 336 pp.
Steffen, W., et al., 2015. The trajectory of the Anthropocene: The Great Acceleration. Anthropocene Rev. 2, 81–98.
Stephenson, M.H., 2013. Returning carbon to nature: coal, carbon capture, and storage. Elsevier, Amsterdam, Netherlands. 143pp.
Stephenson, M.H., 2015a. Shale gas in North America and Europe. Energy Sci. Eng. 4, 4–13.
Stephenson, M.H., 2015b. Shale Gas and Fracking: The Science Behind the Controversy. Elsevier, Amsterdam 153pp.
The Industry of Nations, as exemplified in the Great Exhibition of 1851. Part 2. A Survey of the Existing State of Arts, Machines and Manufactures. Christian Knowledge Society, London, 1852-55.
Waters, C.N., et al., 2016. The Anthropocene is functionally and stratigraphically distinct from the Holocene. Science 351, 137–147.

CHAPTER 4

The Coming Industrial Revolution? Fossil Fuels and Developing Countries

Contents

In the 18th, 19th and 20th centuries, the countries of northern Europe and North America used their fossil fuels to grow and to industrialise, in the process making people generally wealthier and healthier, but also degrading the environment and putting large amounts of geological carbon into the atmosphere. In the present-day less well-developed countries of the south, the same potential lies in fossil fuels—the resources of coal in India and South Africa are for example very large—and the same needs exist to grow and provide wealth and better health. In India, for example, the government would like to move its citizens away from cooking with natural fuels like dung and wood and to stimulate industry, for example in rural areas. South Africa's vast mining potential for metals critical for modern renewables technology will not be realised without power. The question is whether developing countries in their quest for wealth and better health will use their coal, gas, or oil, or bypass the fossil economy straight to a renewables economy.

The purpose of the previous chapter was to illustrate the rise of fossil fuels in the fossil economy of the developed world and in the industrial revolution, the rise of oil, and now the rise of gas. The use of fossil fuels and the use of energy have been, and still are, closely related to population and economic growth, the generation of wealth, and ultimately to poverty alleviation.

Energy and Climate Change
https://doi.org/10.1016/B978-0-12-812021-7.00004-X

The subject of this chapter is to examine whether the same dash for fossil fuels will occur in the developing world in its pursuit of economic growth.

WHERE ARE THE RESOURCES AND RESERVES?

First, it is relevant to establish the distribution of resources. The three maps (Fig. 4.1) show the locations of the three major fossil fuels, their abundance and the relative amounts of reserves and resources. Taking North American oil first, the distinctive feature of its fossil-fuel spectrum is that it is dominated by unconventional oil, which can only be extracted by hydraulic fracturing. The region with the biggest proven reserves of oil is the Middle East. The main geological reason for this region having so much oil is the relative stability of the Arabian (or Persian) Gulf during the last 500 million years of geological history, which has meant that an enormous thickness of sedimentary rock has built up containing layers capable of generating oil and layers capable of storing it. This great pile of sedimentary rock also has the right structures to trap the oil as it ascends. In gas, the Middle East and Russia dominate proven reserves.

Coal has a different profile in the sense that, worldwide, resources are much larger than reserves because coal is known to be present but some is inaccessible or uneconomical to mine. North America, North Asia and South Asia, and Australasia are all well-endowed with coal.

Apart from coal, the main areas of the developing world are not particularly well endowed with fossil fuels, though it is likely that they may also be underexplored for these resources. Of course, not having fossil fuels does not mean these countries will not want to use them, because they will be able to import them.

What do economic forecasters say about fossil-fuel usage in the developing world and elsewhere?

USE OF THESE FOSSIL-FUEL RESOURCES IN THE FUTURE

At least three organisations provide forecasts of fossil fuel use: the International Energy Agency (IEA), the United States Energy Information Administration (EIA), and the oil company BP. They publish regular updates on where they think energy will come from and be used into the future. The most detailed forecasts are those of the IEA.

The IEA was established by the Organization for Economic Cooperation and Development (OECD) in 1974 after the 1973 oil crisis. It forecasts using three 'scenarios' which contain predictions of energy infrastructure

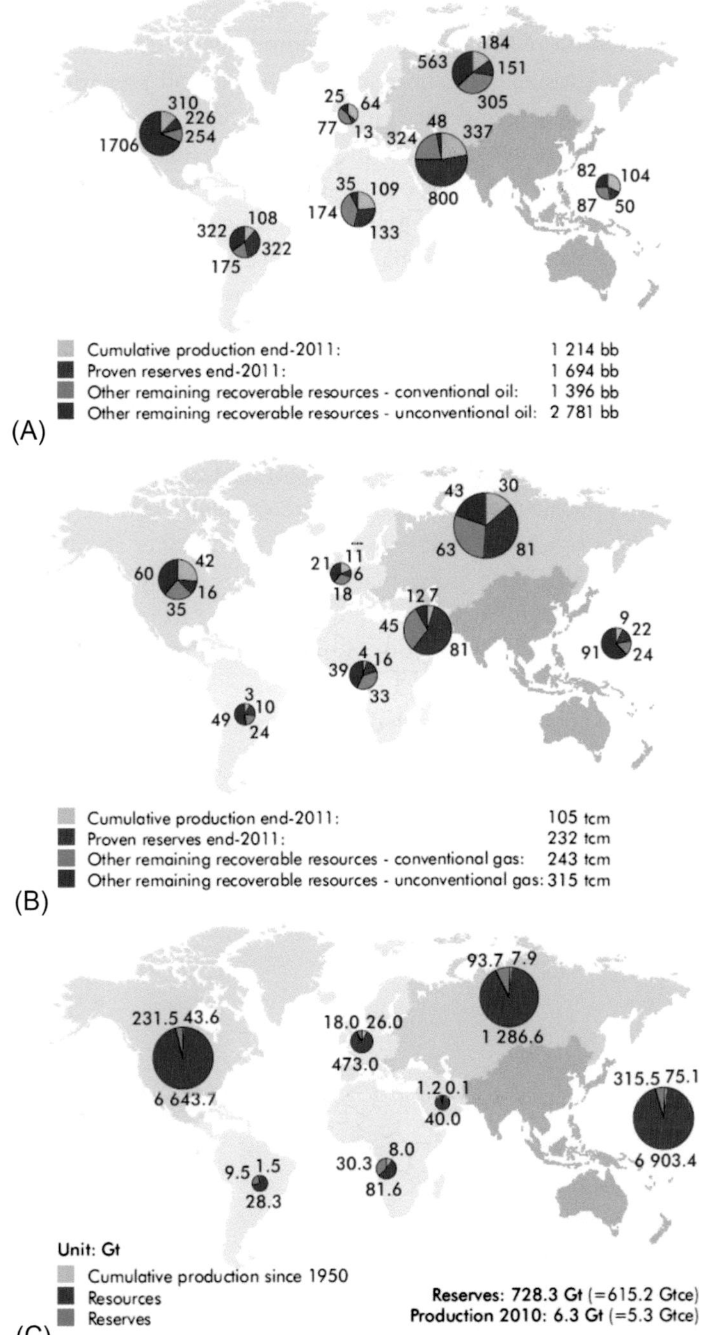

Fig. 4.1 Global resources and reserves of three fossil fuels. (A) oil, (B) gas, (C) coal. *(From https://www.iea.org/publications/freepublications/publication/Resources2013.pdf.)*

investment, energy demand, and supply, based primarily on the need to reduce CO_2 emissions and on common assumptions of economic conditions and population growth. The three scenarios—the '450 Scenario', the 'Current Policies Scenario', and the 'New Policies Scenario'—differ mainly in how various government policies might be applied. In terms of limiting CO_2 emissions, the 450 Scenario is the most stringent in that its projections are constrained by the need to limit concentration of greenhouse gases in the atmosphere to around 450 ppm. This concentration of CO_2 is widely believed to be required to limit global increase in temperature to 2°C. On the other side of the coin is the Current Policies Scenario which looks at present policies in energy assuming that they will not change and calculates the future from this standpoint, regardless of the climate consequences. For example, the IEA's 2016 World Energy Outlook Current Policies Scenario contains only policies enacted as of mid-2016. It is similar to the *business as usual* projections used by climate-change scientists.

The third is the New Policies Scenario, which takes account of broad policy commitments and plans that have been announced by countries and their governments, including national pledges to reduce greenhouse-gas emissions and plans to phase out fossil-energy subsidies, even if the measures to implement these commitments have yet to be identified or announced. This might be regarded as the most realistic and widely quoted of the IEA's scenarios.

The IEA's 2016 World Energy Outlook New Policies Scenario predicts a few interesting trends including, not surprisingly, an increase in energy demand between now and 2040 (Fig. 4.2); 30% of this increase will be from developing countries, particularly in Asia and Africa.

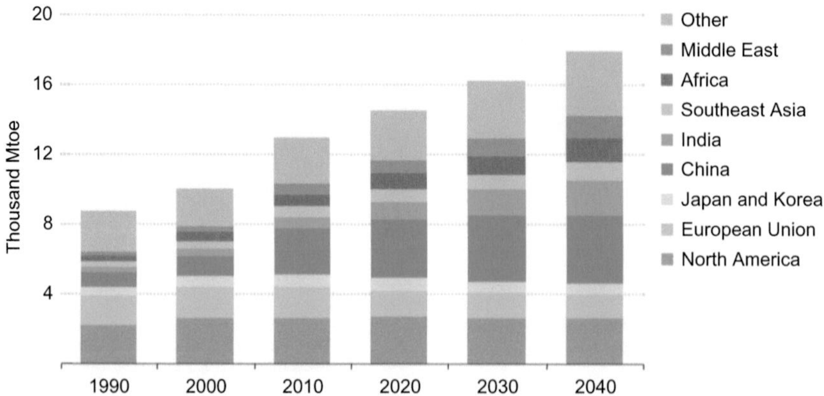

Fig. 4.2 The changing geography of world energy demand. *(From International Energy Agency 2016. World Energy Outlook, 684pp.)*

Breaking this down into types of energy, including the three major fossil fuels, shows an interesting story of surprisingly little change between now and 2040 (Fig. 4.3). Important factors to note include the marked increase in renewables, but also that all three of the main fossil fuels do not decline but in fact *increase*.

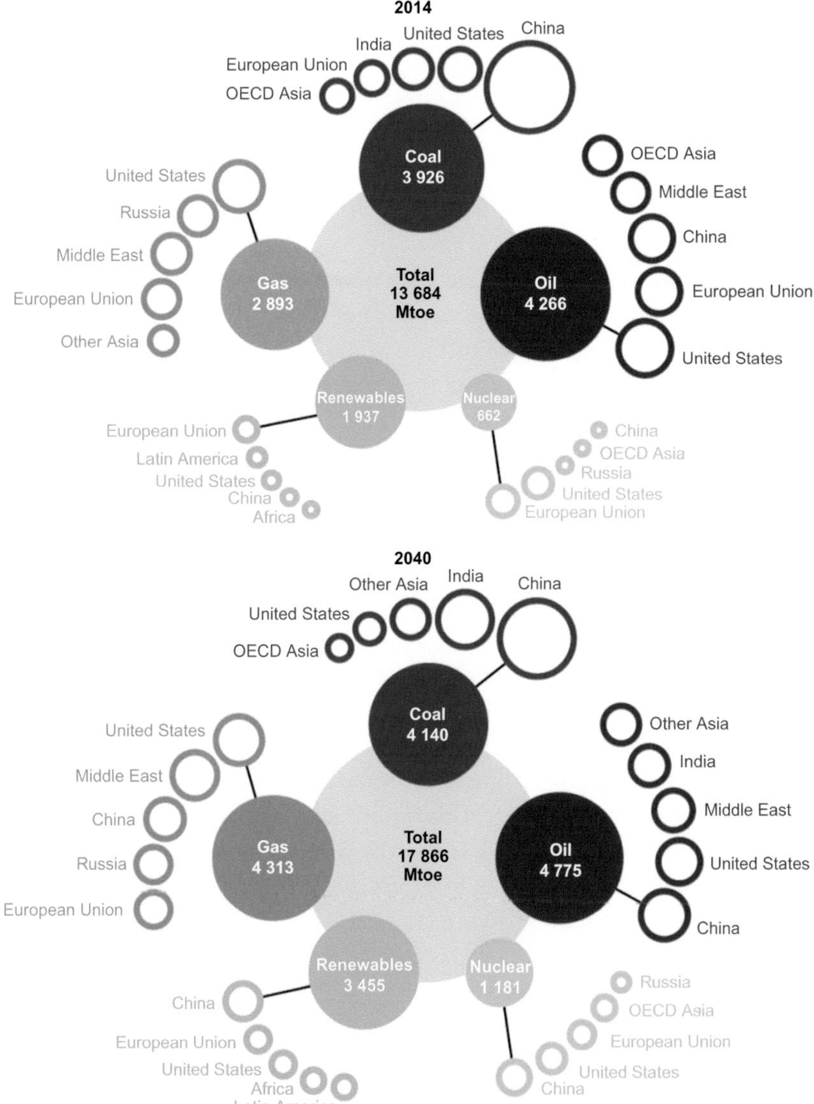

Fig. 4.3 Global primary energy mix, and change between 2014 and 2040. Note the marked increase in renewables. *(From International Energy Agency 2016. World Energy Outlook, 684pp.)*

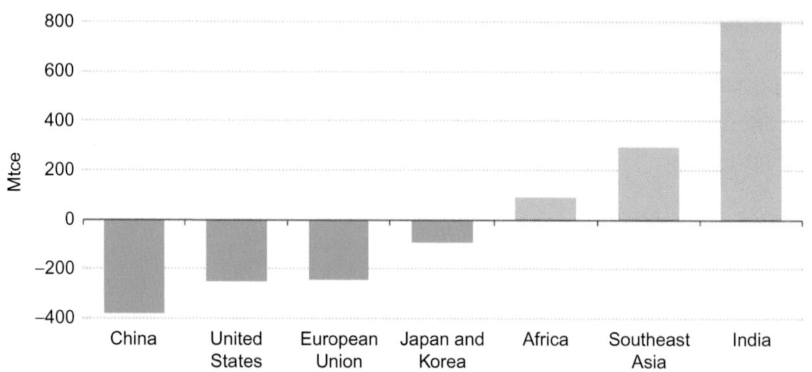

Fig. 4.4 Change in coal demand by key region in the New Policies Scenario. *(From the International Energy Agency 2016. World Energy Outlook, 684pp.)*

Focusing on coal alone in the IEA's New Policies Scenario reveals the shift from the developed to the developing world (Fig. 4.4).

The International Energy Outlook of the EIA, the IEA's American rival, reports similar persistence of fossil fuels in their Reference Case projection. This forecast takes into account known technology and technological and demographic trends and assumes that current laws and regulations are maintained throughout the projections, and so is perhaps closest in spirit to the 'Current Policies Scenario' of the IEA. The EIA forecasts that even though consumption of non-fossil fuels is expected to grow faster than consumption of fossil fuels, fossil fuels still account for 78% of energy use in 2040. According to the EIA, natural gas will be the fastest-growing fossil fuel with consumption increasing by 1.9% per year, including rising supplies of shale gas and coalbed methane. The EIA also says that, although oil will remain the largest source of world energy, its consumption will fall from 33% around now to 30% in 2040. According to the same forecast, coal will be the world's slowest-growing energy source, rising by only 0.6% per year.

BP's 2017 Energy Outlook is broadly similar, though generally it looks out to 2035 rather than 2040. It forecasts the same dominance for fossil fuels in world energy but with growing renewables—and gas growing faster than oil and coal.

It is worth looking at why particular fossil fuels are set to increase rather than decrease in some developing countries, using some examples.

Coal

Coal is still seen by many countries as a viable option as an energy source, because it is very abundant, and, if transport is kept at a minimum, it is cheap.

In India coal is related—through rural electrification—to poverty alleviation and improved health. India has a third of the world's poor. Amongst these poor Indians, who live mostly in rural areas, domestic use of electricity is rare. India alone accounts for more than 35% of the world's population without electricity access. Most domestic energy in India comes from biomass (firewood, crop residue, dung) and India consumes 200 Mtoe (million tonnes of oil equivalent) of biomass each year. One hundred million Indian households still use firewood to cook food, mainly in rural areas. In developing countries generally, many people still rely on biomass for cooking (Fig. 4.5).

Cooking with firewood takes its toll on the health of Indians, with an estimated 50,000 deaths per year (household fires, accidents and ill health). Worldwide, exposure to smoke emissions from the household use of solid fuels is estimated to result in 1.3 million deaths annually. Illnesses are also caused by domestic lighting using kerosene lamps—which is the most common type of lighting in rural India.

An aim of India's rural electrification programme is therefore to introduce healthier fuel in households. Another aim of the programme is to improve agricultural production (through for example better irrigation pumps), and to develop business and trade in agriculture. Partly due to successful electrification, India adds 40 million people to its middle class every year.

Where will India's new electricity come from? The forecasts of the IEA, EIA and BP suggest that most will come from coal. At present, coal provides

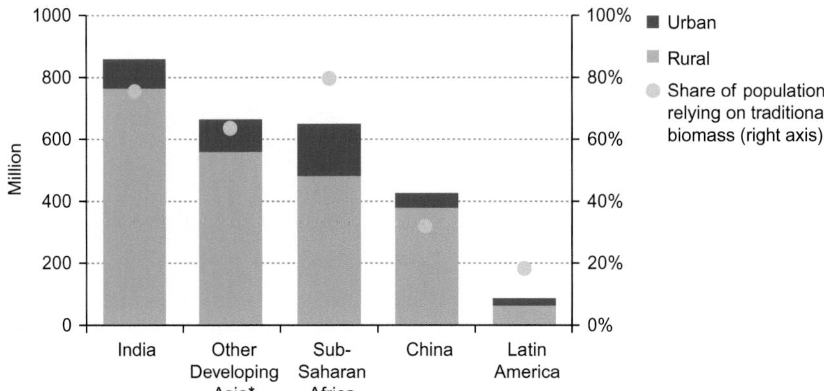

Fig. 4.5 Number and share of population relying on the traditional use of biomass as their primary cooking fuel in 2009. *Includes developing Asian countries except India and China. *(From Kaygusuz, K., 2012. Energy for sustainable development: A case of developing countries. Renew. Sustain. Energy Rev. 16, 1116–1126.)*

about 70% of India's electricity but about 243 GW of coal-fired power is planned in India, with 65 GW actually being constructed and an extra 178 GW proposed. Work led by Christine Shearer of the charity CoalSwarm has surveyed this proposed 'fleet' of coal power stations to forecast the amounts of power that could be provided should these power stations be completed (Fig. 4.6).

Their survey shows average annual capacity additions beginning in 1960, as well as future additions based on proposed new plants. Between 1960 and 2006 there was slow growth but in 2007 government policy changed and the market became more open to the private sector. Between 2011 and 2015, India added 15–22 GW of coal power to its network each year. For the future, Shearer's survey shows that coal plants under development could be producing 435 GW of coal power by 2025, and, assuming an average

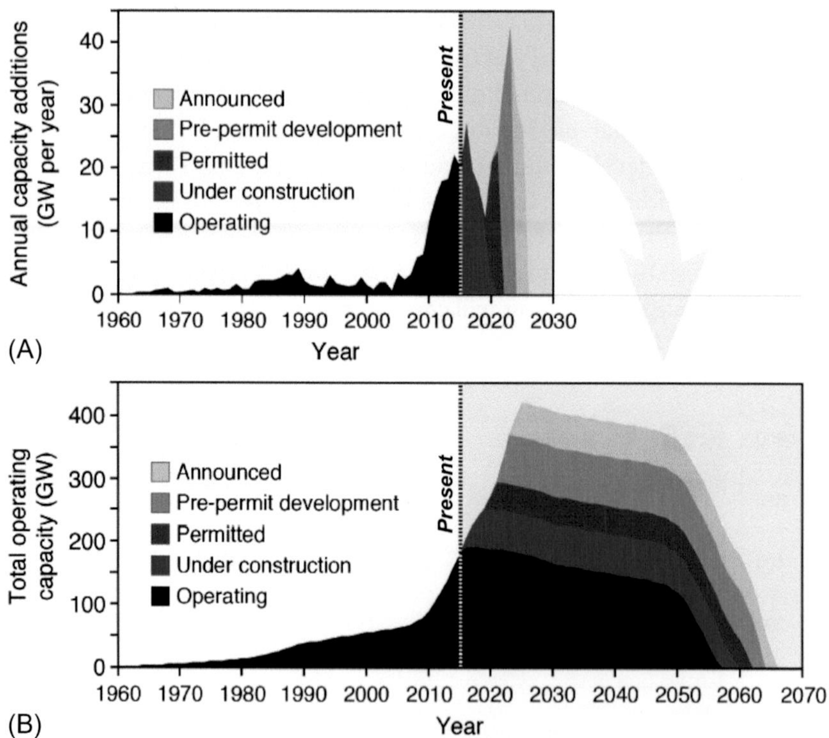

Fig. 4.6 Forecast of coal power station capacity additions in India. (A) Annual capacity additions; (B) total operating capacity. *(From Shearer, C. et al., 2017. Future CO_2 emissions and electricity generation from proposed coal-fired power plants in India. Earth's Future 5, 408–416.)*

lifetime of 40 years, the coal plants could be operating as far ahead as 2065. Such a commitment to coal would guarantee high Indian greenhouse gas emissions for many years to come and prolong the dominance of fossil fuels, freezing out renewables.

How does India's coal use change according to the IEA's scenarios that were described earlier? In the New Policies Scenario, India will become the world's second largest consumer of coal (after China) by around 2025, with demand almost doubling to 880 million tonnes per year by 2035 (Fig. 4.7).

Although today 300 million Indians lack domestic electricity, by 2030 this will have reduced by half. Also, according to the New Policies Scenario, the demand for coal in industrial use will continue to rise, averaging 4% per year growth up to 2035, mainly through the increased manufacture of crude steel. India is also a large cement manufacturer and much industrial coal is used in the process. There is little doubt that this will grow too as more economic growth further spurs building.

South Africa, like India, has a large number of rural people without access to electricity (roughly 60% of South African households), but also a strong demand for electricity, particularly for the mining industry. South Africa's coal reserves are large—28 billion tonnes—which would allow 100 more years of mining at current rates. South Africa exports coal mainly through the port at Richard's Bay, but constraints on rail transport have reduced its capacity to bring coal from the interior and so export has become less important to the country.

According to the IEA New Policies Scenario, South African coal production will by driven mainly by domestic demand for coal for power

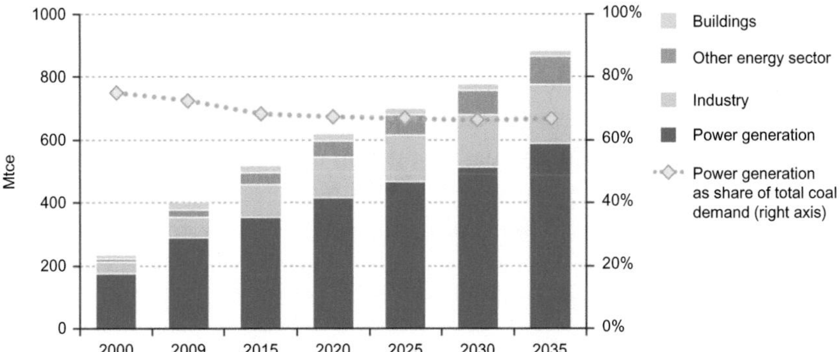

Fig. 4.7 Coal demand in India by sector in the New Policies Scenario. Electricity will be the major user as far ahead as 2035. *(World Energy Outlook 2011 © OECD/IEA 2011, Fig. 10.23, p. 388.)*

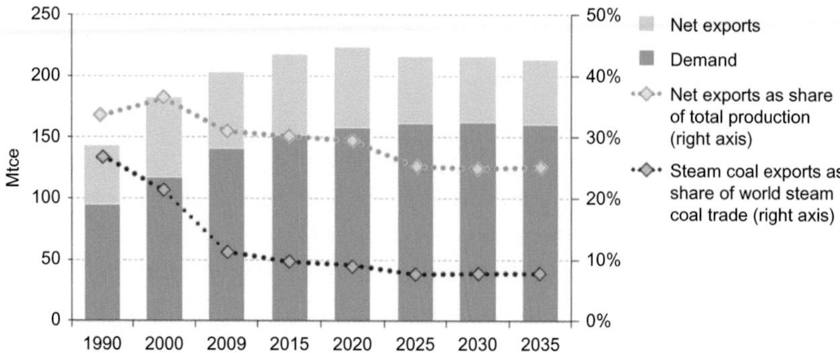

Fig. 4.8 Forecast coal production, exports and demand in South Africa according to the IEA New Policies Scenario. *(World Energy Outlook 2011 © OECD/IEA 2011, Fig. 11.25, p. 441.)*

supply. At present, more than 90% of electricity is generated by coal in South Africa and this will remain the case well into the next decades. Coal production is predicted to rise to a peak of around 230 Mtce around 2020 and then fall to 210 Mtce by 2035 (Fig. 4.8).

Two key factors differentiate South Africa from India. One is the metals mining industry, which is a big user of electricity, particularly in the processing of ore. This industry employs almost half a million South African workers and generates almost a fifth of South Africa's GDP. Another difference is that South Africa also uses a lot of domestic coal in coal-to-liquids (CTL). Sasol, the biggest CTL company in the world, produces 36% of liquid fuels consumed in South Africa, mainly from coal.

Gas

Gas is perhaps more interesting than coal in that it is forecast by the IEA, EIA and BP to grow conspicuously faster than the other two fossil fuels, such that some commentators forecast gas to be the 'fuel of the 21st century'. As I will show later in this section this 'dash for gas' in developed countries is underway and the reasons for it are clear. The IEA New Policies Scenario forecasts gas demand to grow to 2040 in Africa and India, perhaps for similar underlying reasons.

Why is gas the favoured fossil fuel? This probably relates to its relatively low carbon emissions when burnt, its abundance, the development of shale gas, and the revolution that is taking place in the ability to move large amounts of gas around by sea in the form of liquefied natural gas (LNG). The roots of this shift to gas are already visible in developed countries that have a challenge in intermittency risk related to the rise in renewables electricity.

A position paper, 'Flexibility concepts for the German power supply in 2050' led by the German National Academy of Sciences, showed various scenarios to achieve 80%, 90% and 100% reductions in German CO_2 emissions by 2050 (compared with 1990). The 80% scenario is key because it is in line with the EU low-carbon economy roadmap and with legally binding targets set by countries such as the United Kingdom, through its Climate Change Act.

In the German 80% option, the majority of Germany's power for 2050 comes from photovoltaics and wind (Fig. 4.9). The other big share is from gas-powered combined cycle gas turbines (CCGT). It is interesting to note that they also see gas as providing a fill-in for the intermittency of renewables, where power plants adjust their output as demand for electricity fluctuates throughout the day. This is known as 'load following'. In the United States, many gas plants already operate in load-following mode as a backup for renewables, for example in Texas where gas plants fill in for wind.

There are two interesting implications for this. One is the difficulty and cost of operating gas plants when they are not needed at full capacity. The other is predicting as precisely as possible when they will be needed, given the vagaries of the weather (wind and sun), at least a few days in advance.

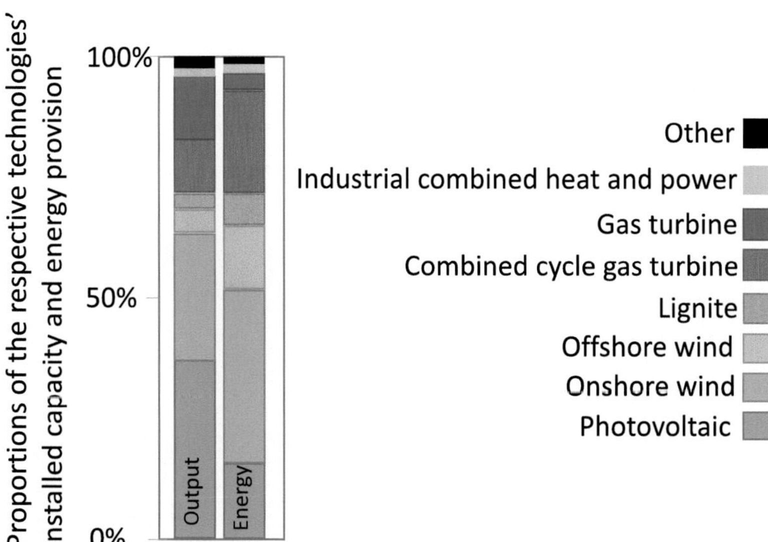

Fig. 4.9 Ways of reaching reduced emissions in German power generation. In the 80% option, the majority of Germany's power for 2050 comes from photovoltaics and wind, and the other big share is from gas-powered combined-cycle gas turbines. *(Redrawn from German National Academy of Sciences Leopoldina, 2016. Flexibility concepts for the German power supply in 2050, 60pp., Fig. 7.)*

In the UK, the Energy Technologies Institute (ETI) advocates the use of hydrogen in the load-following system. The idea is to manufacture hydrogen from methane in a steam reformer running continuously, feeding hydrogen to a storage vessel when grid power is not needed. The stored hydrogen is then fed to the turbine when power is needed. This makes load following more efficient. Fitting the theme of this book, using hydrogen in this way also has a geological aspect in that large-scale hydrogen storage, large enough to contribute to grid power, probably requires storage in geological formations in the subsurface. The likely rock candidate for this kind of storage is rock salt, which is effectively impermeable to gases, even gases with such small molecules as hydrogen. This is discussed in a later chapter.

The recent history of UK power generation shows one significant trend—the appearance of gas in the early 1990s and its continued rise as an important contributor to UK electricity (Fig. 4.10).

The reasons for this are the closing down of coal power stations which do not meet stringent emissions targets, and the slow take-up of new nuclear power stations. The fact that gas power stations are also relatively cheap and quick to build adds to their appeal.

In the United States, the EIA forecasts that natural gas-fired generation will shortly exceed that of coal. Natural gas generation first overtook coal on

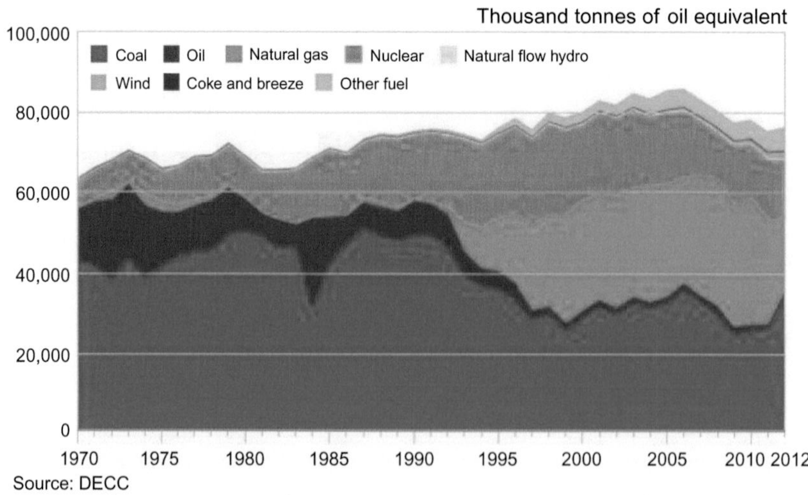

Fig. 4.10 Sources of UK electricity. *(From BBC DECC figures. From http://www.bbc.co.uk/news/business-24823641.)*

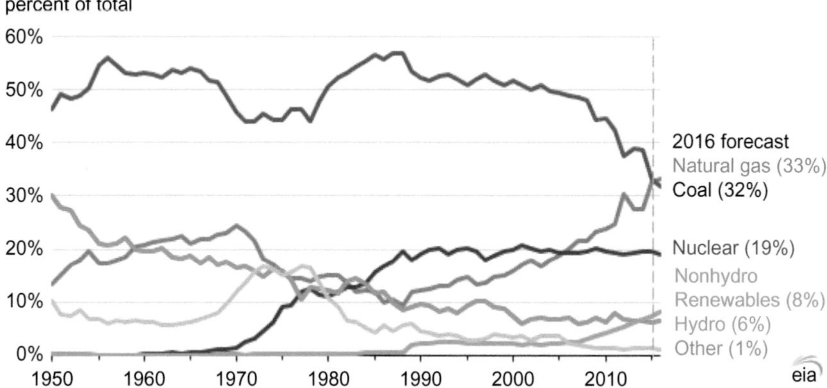

Fig. 4.11 Sources of US electricity. *(From US Energy Information Administration http://www.eia.gov/todayinenergy/detail.php?id=25392.)*

a monthly basis in April 2015. The reason was low natural gas prices beginning in 2009, mainly due to large amounts of shale gas on the market (Fig. 4.11).

Oil

The New Policies Scenario of the IEA forecasts that India will be the largest source of oil demand growth to 2040 in the world, partly due to industrial and social developments including personal car ownership. Africa in general also shows demand in growth.

CAN THE DEVELOPING WORLD 'LEAP-FROG' FOSSIL FUELS?

It is clear from forecasts that many agencies think that the developing world will take up fossil fuels in the years to come, though the shifts between fossil fuels, say from coal to gas, are more difficult to predict. The reasons that many developed countries are shifting to gas may not apply in the future in developing countries.

But are these agencies wrong in assuming a shift to fossil fuels? Recent publications by organisations such as the Africa Progress Panel and the International Renewable Energy Agency (IRENA) advocate a greater focus on renewables, particularly for Africa. They suggest that just as Africa jumped from virtually no telephone communications to a well-developed mobile telephone system without developing landlines, so it could 'leapfrog' to full renewables without developing a comprehensive fossil-fuel economy.

According to reports by IRENA, renewable resource in Africa is very large, with a potential for 300 million MW of solar photovoltaic power and more than 250 million MW of wind (Fig. 4.12).

IRENA and the Africa Progress Panel note the challenges to realise this resource potential. Mostly this means huge investment. Meeting sub-Saharan Africa's power needs will cost US$40.8 billion a year, equivalent to 6.35% of Africa's gross domestic product, according to the World Bank. The locations of wind and solar resources in Africa are not known in enough

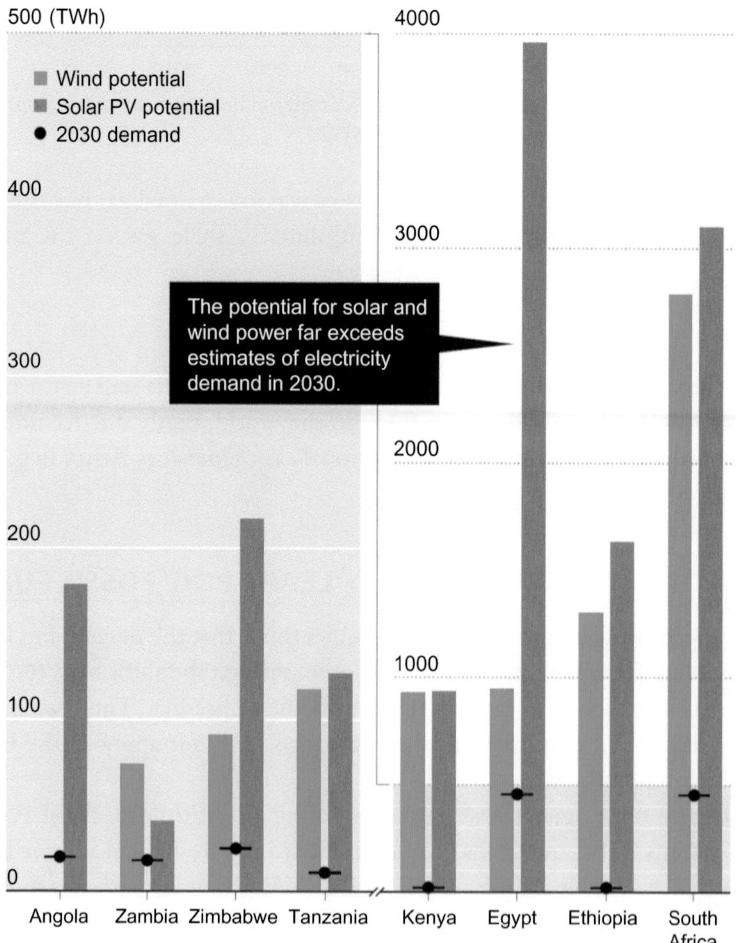

Fig. 4.12 Wind and solar potential in selected African countries including estimate of electricity demand in 2030. *(From Gies, E., 2016. Can wind and solar fuel Africa's future? Nature 539, 20–22.)*

detail at the moment to stimulate private investment by companies hoping to select sites for projects. Another problem is that Africa lacks big electricity grids and transmission lines to move large amounts of power within countries and across regions (Fig. 4.13).

It may be possible to scale up renewable projects and manage electricity better than it is in developed countries. For example, hydroelectric power could be used to fill in for solar and wind intermittency. An Africa-wide grid could also smooth intermittency effects because solar and wind are likely to be feeding into the grid at least some places on the continent at any one time. Smart grids could develop demand response, which reduces electricity

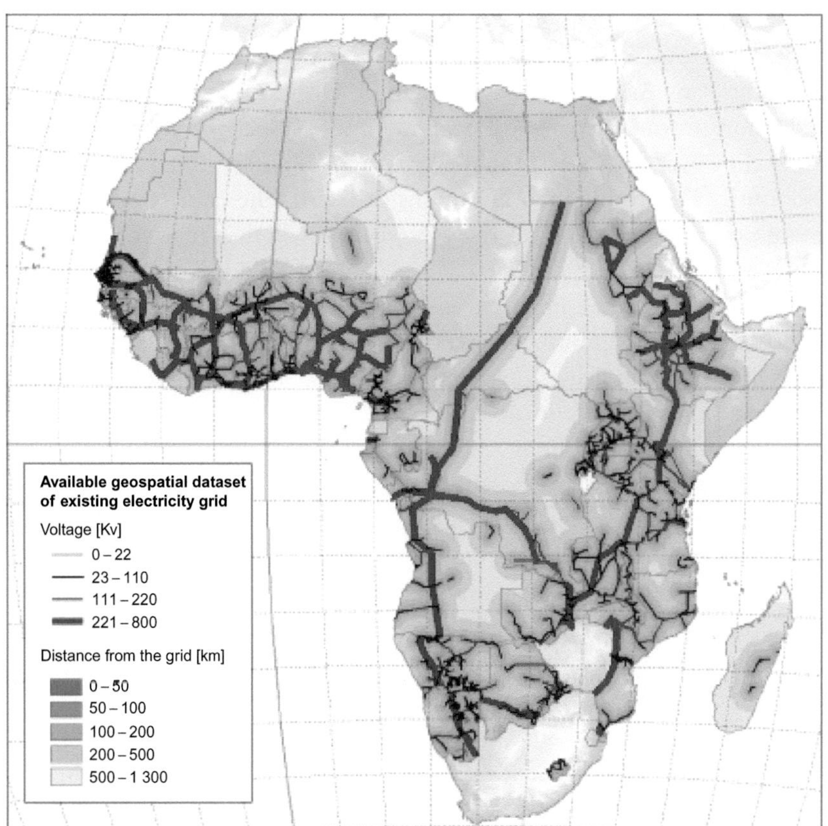

Fig. 4.13 Available geospatial dataset of existing electricity grid in Sub-Saharan Africa and distance from the grid. *(From Szabo, S., 2011. Energy solutions in rural Africa: mapping electrification costs of distributed solar and diesel generation versus grid extension. Environ. Res. Lett. 6 (034002), 9pp.)*

delivery to multiple customers by imperceptible amounts when demand is peaking.

But coming climate change could be a problem for hydroelectric power in that, in many parts of Africa, river flows are forecast to decline in long-term droughts. Perhaps there would be a role for gas in the same way that developed countries see a role for gas.

THE 'PARIS AGREEMENT' AND THE 'TWO-DEGREE WORLD'

The United Nations Framework Convention on Climate Change (UNFCCC) is an international treaty agreed upon at the Earth Summit in Rio de Janeiro in 1992. The objective of the UNFCCC is to 'stabilize greenhouse gas concentrations in the atmosphere at a level that would prevent dangerous anthropogenic interference with the climate system'. In 2015 the Paris Agreement was adopted by UNFCCC, to govern emission reductions from 2020 onward, and countries publicly outlined what post-2020 climate actions they would take under the agreement, known as their Nationally Determined Contributions (NDCs). The climate actions communicated in these NDCs largely determine whether the world achieves the long-term goals of the Paris Agreement: to hold the increase in global mean surface temperature to well below 2°C above preindustrial levels, to pursue efforts to limit the increase to 1.5°C, and to achieve net zero emissions in the second half of this century. The Paris Agreement is the world's first widely supported climate agreement.

Recent research by Joeri Rogelj and colleagues seems to show that current emission pledges as part of the Paris Agreement are not sufficient to keep temperature increase below 2°C above preindustrial levels. The NDCs collectively lower greenhouse gas emissions, but still add up to a median warming of 2.6–3.1°C by 2100. However, the Paris Agreement requires that emission controls become more stringent over time. According to the same researchers a '...substantial enhancement or over-delivery on current NDCs by additional national, sub-national and non-state actions is required to maintain a reasonable chance of meeting the target of keeping warming well below 2 degrees Celsius'.

The questions are: how can the NDCs be more effective and how do we check that they deliver what they promise?

There are as yet few details on how to track progress in the Paris Agreement and the NDCs. Early assessments of progress point to a recent slowdown in global emissions growth due to reduced growth in coal use since

2011 in China and the United States, partly due to economic slowdown and the advent of shale gas. Growth in wind and solar has contributed to the global emissions slowdown, but has been less important than economics and energy efficiency.

How do the IEA scenarios relate to the aims of the Paris Agreement? The New Policies Scenario does not intrinsically contain all the policies of the NDCs and the concept that the NDCs will tighten as required as time goes by, but tries to judge the efficacy of the NDCs to deliver the change needed. As the IEA's World Energy Outlook of 2016 says: '...The extent to which countries will eventually deliver on their pledges in the energy sector critically depends on two main factors: the policies that support the required longer term structural energy sector transition, and the short-term macroeconomic and energy market trends, which may accelerate—or impede—the transition towards a lower carbon energy future...'.

This cautious approach is shared by climate commentator Kevin Anderson, who in 2015 published an article in the journal *Nature* expressing concern that the NDCs rely too heavily on unproven technologies, particularly on the so-called negative-emission technology BECCS (bioenergy and carbon capture and storage). BECCS involves the burning of biomass (rather than fossil fuels) and the subsequent sequestration of resultant CO_2 in deep underground stores. I will examine the pros and cons of BECCS and other carbon sequestration technologies in the next chapter.

SUMMARY

Economic forecasts suggest energy demand will increase in the developing world. This is not a surprise; however, the extent in the forecasts to which this demand will be satisfied by fossil fuels is worrying and appears to bear out the possibility that another industrial revolution involving the fossil economy might be replayed in the cities and countryside of India, other parts of Asia, and Africa. The fossil economy allowed increasing population in the developed world, and made the developed world rich. It is also linked to exploitation and of course to environmental degradation beyond that caused by greenhouse gas emissions.

It is difficult to deny the developing world its chance to become rich, but with the fossil economy's wealth come emissions. Most of the increased CO_2 in the atmosphere between the beginning of the fossil economy and now is the responsibility of the developed world.

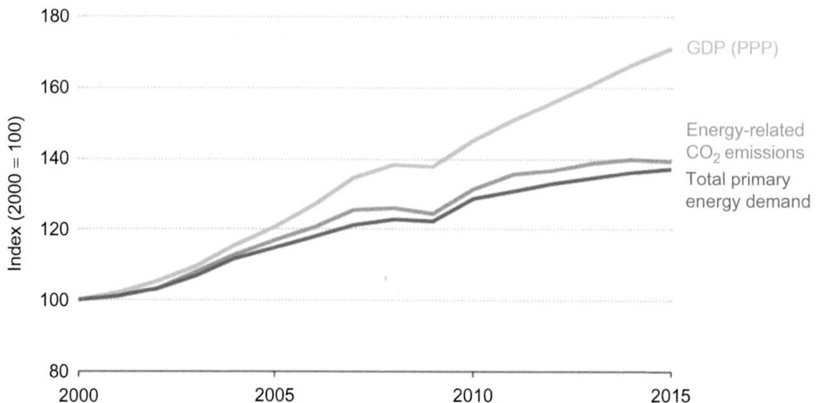

Fig. 4.14 Decoupling of global economic output, energy demand, and energy-related CO_2 emissions, mainly due to the increase in renewables. *(From the International Energy Agency 2016. World Energy Outlook, 684pp.)*

More recently, economic growth has begun slightly to decouple from emissions, mainly due to the increase in renewables (Fig. 4.14). In Africa and India it may be that a full fossil economy could be bypassed through rapid take-up of renewables, though a sophisticated grid will be needed to do without fossil-fuel load following. The cost could be very large but then the investment to improve grids for electricity of the fossil variety also requires a grid. A large-scale renewables grid is of course a more sustainable option. Given that some of the materials to support renewables (e.g., rare-earth metals in wind turbines) will be extracted in the mines of the developing world, it would make sense to develop a home-grown industry for local skills, jobs, and innovation.

BIBLIOGRAPHY

Africa Progress Panel, 2015. Power People Planet: Seizing Africa's Energy and Climate Opportunities. 182pp.

Anderson, K., 2015a. Duality in climate science. Nature Geosci. 8, 898–900.

Anderson, K., 2015b. Talks in the city of light generate more heat. Nature 528, 437.

Bhattacharyya, S.C., 2006. Energy access problem of the poor in India: Is rural electrification a remedy? Energy Policy 34, 3387–3397.

BP Energy Outlook 2017. https://www.bp.com/content/dam/bp/pdf/energy-economics/energy-outlook-2017/bp-energy-outlook-2017.pdf.

Castaneda, C.J., 2004. History of Natural Gas. In: Encyclopedia of Energy. 4, Elsevier Inc., pp. 207–218

Energy Information Administration, 2015. Annual Energy Outlook. 154pp.

Energy Information Administration, 2016. International Energy Outlook. 290pp.
German National Academy of Sciences Leopoldina, 2016. Flexibility concepts for the German power supply in 2050. 60pp.
Gies, E., 2016. Can wind and solar fuel Africa's future? Nature 539, 20–22.
International Energy Agency, 2011. World Energy Outlook. 450pp.
International Energy Agency, 2016. World Energy Outlook. 684pp.
International Energy Agency, 2013. Resources to Reserves: Gas and Coal Technologies for the Energy Markets of the Future. 272pp.
Karl, T.L., 2004. Oil-Led Development: Social, Political, and Economic Consequences. In: Encyclopedia of Energy. 4, Elsevier Inc., pp. 661–672
Kaygusuz, K., 2012. Energy for sustainable development: A case of developing countries. Renew. Sustain. Energy Rev. 16, 1116–1126.
Miketa, A., Saadi, N., 2015. Africa Power Sector: Planning and Prospects for Renewable Energy IRENA. 44pp.
Murthy, N.S., et al., 1997. Economic growth, energy demand and carbon dioxide emissions in India: 1990–2020. Environ. Develop. Econ. 2, 173–193.
Ness, G.D., 2004. Population Growth and Energy. Encyclopedia of Energy. 5, Elsevier Inc., pp. 107–116
Peters, G.P., et al., 2017. Key indicators to track current progress and future ambition of the Paris Agreement. Nature Clim. Change 7, 118–123.
Rogelj, J., et al., 2016. Paris Agreement climate proposals need a boost to keep warming well below 2°C. Nature 534, 631–639.
Rural Electrification (India) Corporation, 2012. 43rd Annual Report.
Shearer, C., et al., 2017. Future CO_2 emissions and electricity generation from proposed coal-fired power plants in India. Earth's Future 5, 408–416.
Sheffield, J., 1998. World population growth and the role of annual energy use per capita. Technol. Forecast. Social Change 59, 55–87.
Szabo, S., 2011. Energy solutions in rural Africa: mapping electrification costs of distributed solar and diesel generation versus grid extension. Environ. Res. Lett. 6 (034002). 9pp.
Wu, G.C., et al., 2015. Renewable Energy Zones for the Africa Clean Energy Corridor IRENA/LBNL. 100pp.

CHAPTER 5

Geology and the Reduction of Emissions

Contents

As well as being fundamental to the long-term carbon cycle, energy, and climate change, rocks also provide ways in which climate change can be reduced or mitigated. Layers of sedimentary rocks provide deep potential repositories for carbon dioxide and through the technology of carbon capture and storage (CCS), the long-term carbon cycle can essentially be speeded up to permanently remove industrial CO_2—associated with power stations and industry—from the atmosphere. Though great strides are being made to decarbonise power generation through the uptake of renewables, there are still few decarbonisation routes for cement and ammonia factories, or refineries. The parts of CCS—CO_2 capture, pipeline transport and geological burial—have been demonstrated separately, but never together on a large scale, in other words at a scale that would make a difference in climate-change terms. A variant on CCS, bioenergy and carbon capture and storage (BECCS), is attractive to politicians and policymakers alike because it offers 'net negative' emissions, though its viability in the face of the required agricultural resources to make it work as an effective large-scale abatement technology is in doubt. More indirect ways that geological materials and resources can reduce emissions or mitigate climate change are through geothermal energy, storage of excess

Energy and Climate Change
https://doi.org/10.1016/B978-0-12-812021-7.00005-1

renewables energy in geological compressed air energy storage (CAES), and geological storage of hydrogen as part of the low-carbon hydrogen economy. Geological materials will likely provide many of the key elements for grid-scale battery storage, and for batteries for electric vehicles. Not to be forgotten is the likely global role that deep geological repositories will play in the safe long-term disposal of waste from low-carbon nuclear power stations.

Although the geological resources that are fossil fuels are responsible for climate-altering emissions, geological materials also offer abundant ways to reduce emissions. I showed in earlier chapters that the long-term carbon cycle not only creates fossil fuels through natural processes, but it can also release emissions from fossil fuels naturally, for example when coal oxidises at the surface of the Earth or when volcanic heat interacts with buried carbon, like coal. Of course, the long-term carbon cycle can also remove CO_2 from the atmosphere through silicate weathering and the formation of coal. Removal of CO_2 from the atmosphere large scale could naturally cool the Earth in a phenomenon known as the 'negative greenhouse effect'.

This chapter looks mainly at a way that the natural processes of the long-term cycle can be speeded up or 'short-circuited' in the technique of geological sequestration, also known as carbon capture and storage (CCS). Some other less direct ways of contributing to low-carbon power are also discussed.

THE GEOLOGICAL PRINCIPLES OF CARBON CAPTURE AND STORAGE

Before we consider how geological sequestration works, it is first necessary to point out that capture of CO_2 from gas mixtures using chemical means is not new; for example, CO_2 has been extracted from the atmosphere in submarines for a long time using soda lime (sodium hydroxide and calcium hydroxide). In commercial CCS the emphasis, though, is on large point sources of CO_2, for example fossil-fuel power stations, ammonia factories, refineries and cement works. The most efficient method at the moment is to bubble waste gas through an amine solvent which absorbs the CO_2. Though expensive, the process is well understood and can provide a steady stream of extracted CO_2 to be disposed of underground.

Moving to sequestration, it may come as a surprise that rocks can be used to store or dispose of any fluid (liquid or gas), because they look impressively solid. As I described earlier, however, sedimentary rocks of a particular kind are made of irregularly shaped particles that have been compressed together with the result that, though they look solid, they actually contain either gas-filled or liquid-filled space.

Sedimentary rocks around the world are known to contain natural gas, but they are also known to contain natural accumulations of CO_2, and more exotic noble gases. In the same way, sedimentary rocks also contain natural accumulations of liquids such as oil; but they are also—and much more commonly—saturated with water below the water table.

Displacement of Pore Fluid by CO_2

The principle of CO_2 disposal or storage in rocks then is simple—to displace the fluid that presently occupies the pores between particles in rock, and fill that space with CO_2. If the rock saturated with CO_2 is deep enough and in the right structure, the CO_2 will stay down under the surface for a very long time, perhaps millions of years. In this way, the CO_2 becomes isolated from the atmosphere and so can no longer act as a greenhouse gas (GHG).

To displace the other fluids in rock will involve drilling a deep hole and then forcefully injecting CO_2 so that other fluids are pushed out of the way. This can be illustrated with a very simple diagram (Fig. 5.1).

The figure shows a large anticline or arch-shaped structure of geological layers. The anticline is composed of an upper layer of impermeable rock like shale or mudstone (a caprock), and a lower layer of porous and permeable rock like sandstone or limestone (a reservoir). A deep borehole penetrates the caprock and provides sufficient pressure for CO_2 to penetrate the borehole, but also leak into the surrounding rock, pushing the other fluids aside. In most subsurface environments this involves pushing saline water out of the way, because at depth this is by far the most common fluid. The process of displacement on a microscopic scale is illustrated in Fig. 5.2.

Of course, sequestration is not as simple as this. A few questions immediately need answering: for example, where does the displaced saline water go, how much CO_2 can be disposed of in total, and what happens to the CO_2 in the long term?

The freer the saline water is to move, the easier it will be to load up the reservoir with CO_2. A reservoir of porous and permeable rock which is extremely large in subsurface extent and which is homogeneously porous and permeable throughout would be ideal. Most reservoir rocks are not like this. They may only extend for a few kilometres either side of the borehole

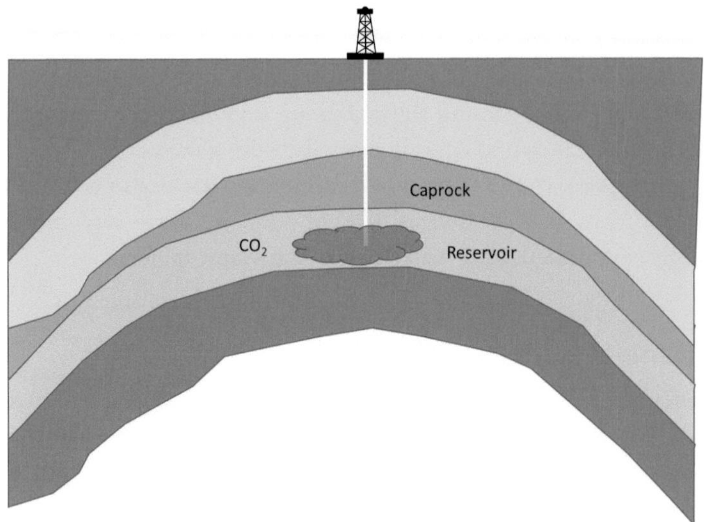

Fig. 5.1 A geological structure that could be appropriate for CCS. CO_2 would be pumped into the pore spaces of the reservoir and held in place by the caprock.

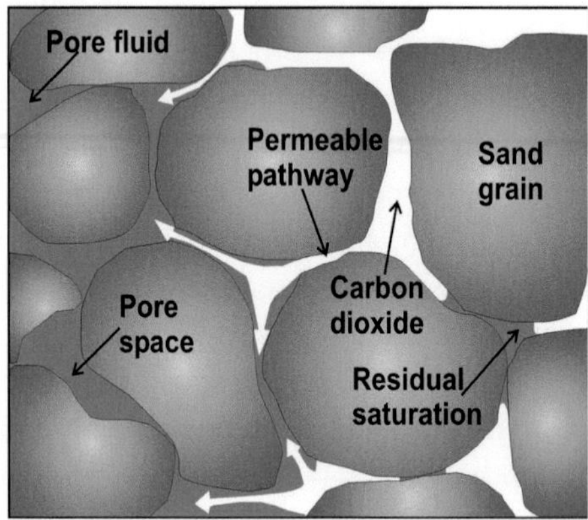

Fig. 5.2 Pumping CO_2 into a reservoir will push the pore fluid (usually saline water) out of the way. Some of the saline water might remain clinging to the sand grains—this is known as residual saturation. *(Courtesy Michelle Bentham. BGS copyright NERC.)*

and may be thin. They may also contain natural internal barriers that impede or completely stop fluid flow.

Fig. 5.3 illustrates one way in which a fluid like saline water could be impeded. It shows a road cutting in Utah with two small geological faults in green. For the purposes of this explanation, two layers of impermeable rock, also in green, have also been imagined.

Fig. 5.3 A road cutting exposing faults in a cliff face. *(Courtesy Sam Holloway and Gary Kirby. BGS copyright NERC.)*

We know from numerous studies that sometimes faults can act as seals, preventing fluid from moving across them. If the upper and lower green layers are impermeable, then the box-shaped area in between is essentially confined in two dimensions, and could be completely confined if other barriers existed outside the plane of the picture.

Injection of CO_2 into this structure would push some of the saline pore water out of the way, but not for very long, because water is not compressible. Very quickly the CO_2 and the water would become highly pressurised in the confined structure, so that the amount of CO_2 injected would be rather small.

In Fig. 5.1 CO_2 is shown collecting immediately under the arch of the anticline. There is a reason for this. Though the CO_2 would be pushing the pore water aside, the CO_2 fluid is also less dense than the pore water and so is buoyant and will collect under the highest part of the structure. This is the principle of the geological 'trap'—a rock structure that holds fluids in place. In this case CO_2 is trapped underneath an impermeable arch. In CCS terminology, this is a 'physical trap' because simple physical processes—buoyancy, flow and pressure—govern it. The same principle applies to oil and gas fields—buoyant hydrocarbons also sit under natural impermeable arches all over the world.

At great depth CO_2 takes up much less space than it does in the atmosphere. This is not just because CO_2 is compressed as a gas with increasing pressure, but also because CO_2 changes phase into a supercritical fluid at about 1 km depth under normal geological conditions. This is a much more dense form than gaseous CO_2 gas, so a lot of CO_2 can be disposed of in relatively little pore space.

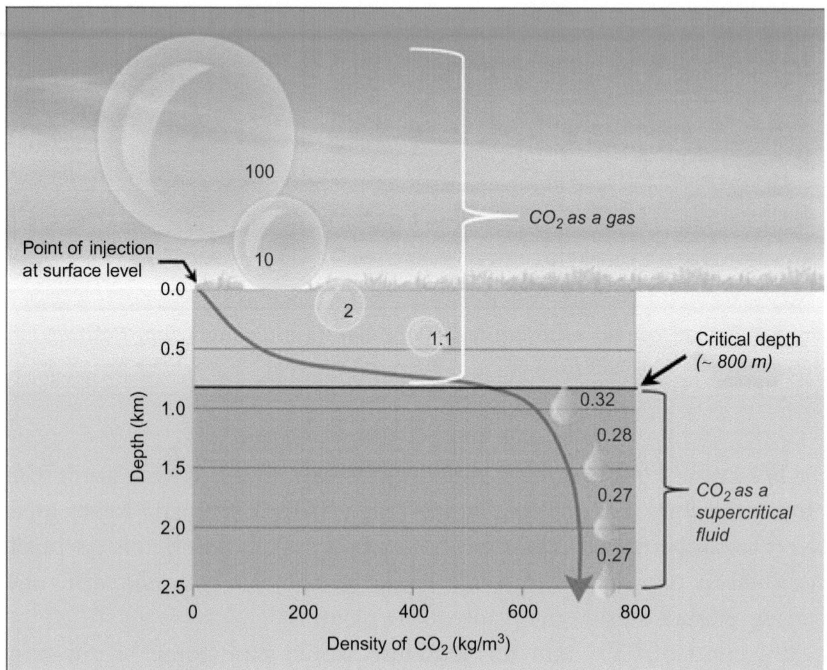

Fig. 5.4 At great depth CO_2 takes up only a quarter of 1% of its original volume at the surface. *(From the NETL website: http://www.netl.doe.gov/technologies/carbon_seq/FAQs/carboncapture3.html.)*

How much space is saved is illustrated in Fig. 5.4. Starting with a CO_2 volume of an arbitrary 100 units, subsurface pressure renders the volume down to only a quarter of one unit—in other words only a quarter of 1% of its original volume at the surface.

Industrial-Sized Storage Space

The UK hopes to dispose of at least some of its industrial CO_2 in a rock layer known as the Bunter Sandstone. Fig. 5.5 shows the sandstone as it appears in a quarry near Scunthorpe in eastern England, and the inset picture shows its microscopic structure. The porosity of the sandstone is 15–20% and the permeability is high.

Though this is the Bunter Sandstone on land, its main potential is as an offshore disposal site in the North Sea, east of the Humber estuary (Fig. 5.6).

The map (Fig. 5.6) shows the extent of the Bunter Sandstone layer both on land and under the seabed. The layer underlies much of England in a Y-shaped pattern, but only outcrops in thin strips around northeastern

Fig. 5.5 The Bunter Sandstone in a quarry near Scunthorpe in eastern England. The inset picture shows the porosity and permeability of the sandstone. *(Courtesy Sam Holloway and Graham Lott. BGS copyright NERC.)*

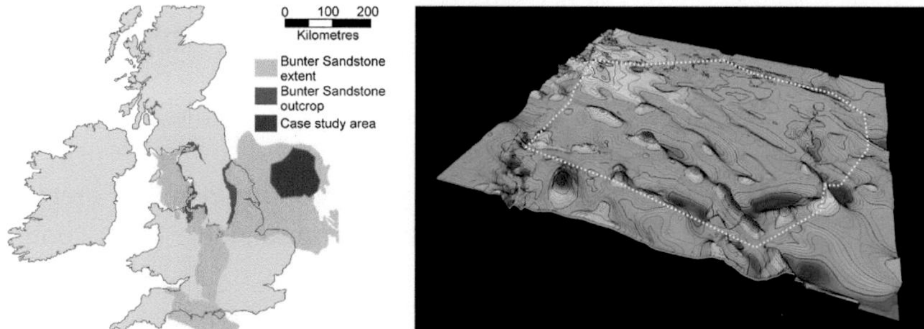

Fig. 5.6 Left. The outcrop and underground extent of the Bunter Sandstone. Much of it is under the seabed of the North Sea. Right. 3D representation of the Bunter Sandstone in part of the North Sea. The white dotted line shows the edge of the 'case study' area shown at left. The area within the white dotted line is about 10,000 km^2. *(Courtesy Andy Chadwick. BGS copyright NERC.)*

England and the Midlands. It reaches below the surface in eastern England and continues offshore. A 3D computer model has been created of the layer in a part of the North Sea and this is shown in Fig. 5.6 in red. The Bunter Sandstone is 200–350 m thick, but over the area of the case study appears thin in the diagram. The colour coding indicates the depth of the layer below seabed, red meaning shallow and blue meaning deep, in fact several kilometres below the seabed.

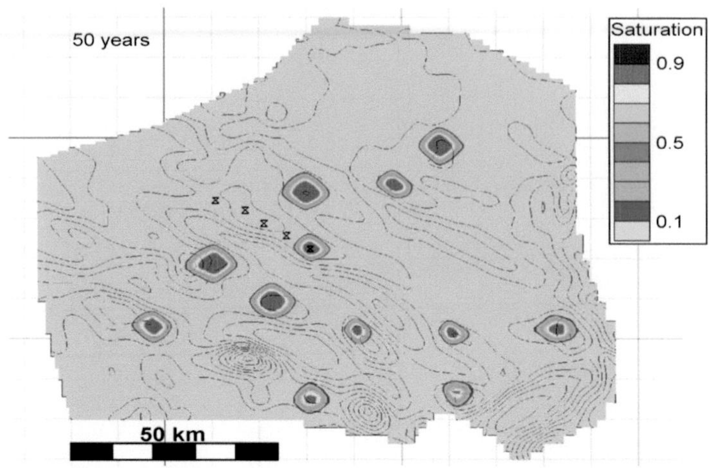

Fig. 5.7 The result of the simulation in the case study area: injecting 33 million tonnes of CO_2 per year for 50 years. *(From Smith, D.J. et al., 2011. The impact of boundary conditions on CO_2 storage capacity estimation in aquifers. Energy Procedia 4, 4828–4834.)*

The injection of CO_2 can be simulated in the case study area to calculate how much of the gas can be stored or disposed of. The simulation imagined 12 injection points in the case study area injecting a combined rate of 33 million tonnes of CO_2 per year for 50 years (Fig. 5.7). This is a rate equivalent to about three large coal-fired power stations.

The colours represent CO_2 saturation or the proportion of the pores that is occupied by CO_2, red representing the highest saturations. The 12 points show the location of accumulated CO_2 over 50 years, a total of 1650 million tonnes, mainly around the injection wells which are themselves located on arch-shaped structures which encourage trapping. The simulation also indicates that less than 1% of the total pore volume of the case study area is occupied by CO_2, and therefore that the Bunter Sandstone layer could easily accommodate 50 years' emissions, equivalent to three large power stations. The simulation indicates crudely that storage or disposal space is on a scale consistent with modern industrial power generation.

Simulations have also been carried out on individual traps to investigate how buoyant supercritical CO_2 behaves after it has been injected into a reservoir. Fig. 5.8 shows a model of a single structure of the Bunter Sandstone simulated for 50 years of CO_2 injection, showing distribution of CO_2 at different periods of time after the start of injection, in fact at 3, 10, 50, 150 and 1781 years.

The simulation reflects the buoyancy of CO_2 seeking the highest position under the impermeable structure that confines it. In the first image

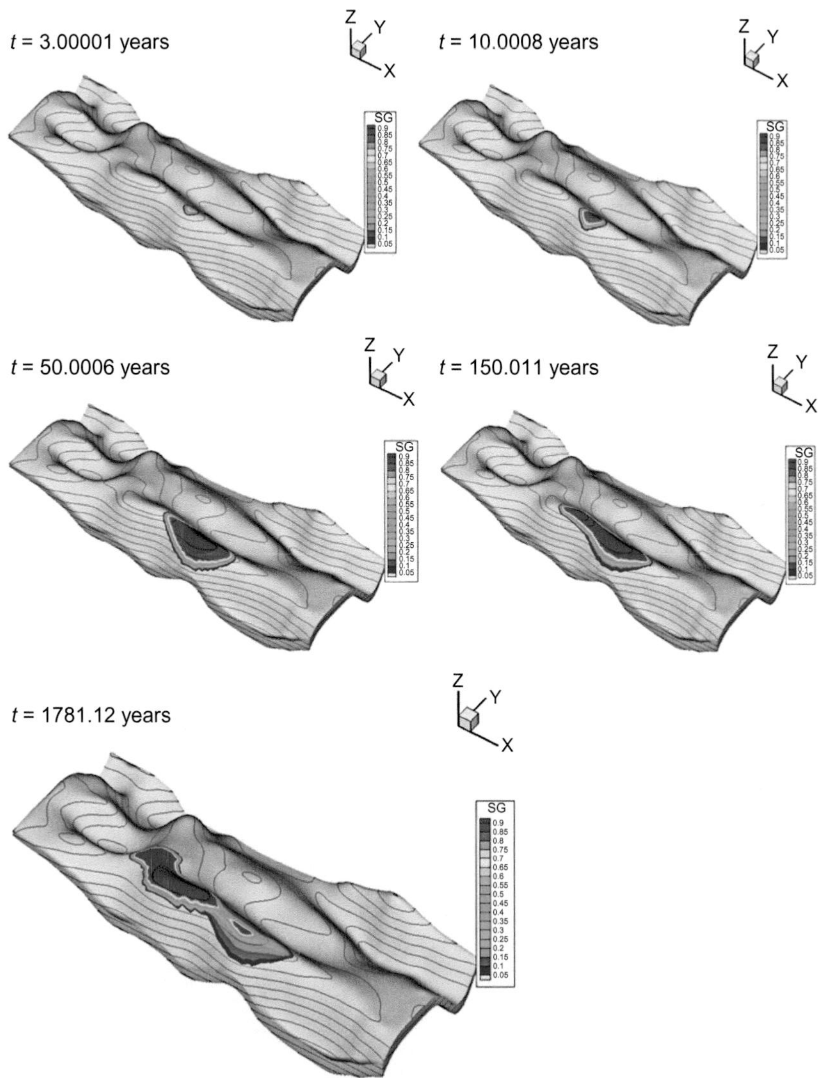

Fig. 5.8 Simulation of buoyant trapping of CO_2. *(Courtesy of David Noy and Andy Chadwick. BGS copyright NERC.)*

($t = 3$ years), injection has just started and so the saturated area of CO_2 is very small. The highest saturation is close to the injection well. In the second picture (10 years), 10 years of accumulated CO_2 forms a circular patch in plan view. At 50 years, the process of buoyancy acts such that the CO_2 migrates up the arch-shaped structure. At 150 years (100 years after the end of injection), much of the CO_2 has moved into the highest parts of the structure. After 1781 years, it has almost left the original arch structure.

The simulations I have described allow estimates for the size of storage space. The *Carbon Sequestration Atlas of the United States and Canada* published by the US National Energy Technology Laboratory (NETL) suggests that there is enough space in rocks in the United States and Canada for 100 years of CO_2 at present rates of emission. A similar figure for Western Europe is estimated by the European Union 'Geocapacity Project'.

In Britain the United Kingdom Storage Appraisal Project (UKSAP) suggested a storage figure of 78 gigatonnes (Gt), in crude terms enough room for thousands of years of CO_2 from Britain's power stations. The diagram (Fig. 5.9) shows how this figure of 78 Gt is broken down. The largest proportion by far (60 Gt), 'nonchalk aquifers', are mainly deep saline aquifers of sandstone. The chalk of southern England could also provide storage space. The depleted fields of oil and gas that I discuss in a later section provide about 10 Gt of storage.

The UKSAP project identified the best storage in the northern, central and southern North Sea, but also in the east Irish Sea and the English Channel (Fig. 5.10).

Long-Term Behaviour of CO_2

The simulations illustrated earlier deal essentially with buoyancy and injection. Over very long periods, other factors become important, for example dissolution of CO_2 in pore water and reaction of CO_2 with the minerals in

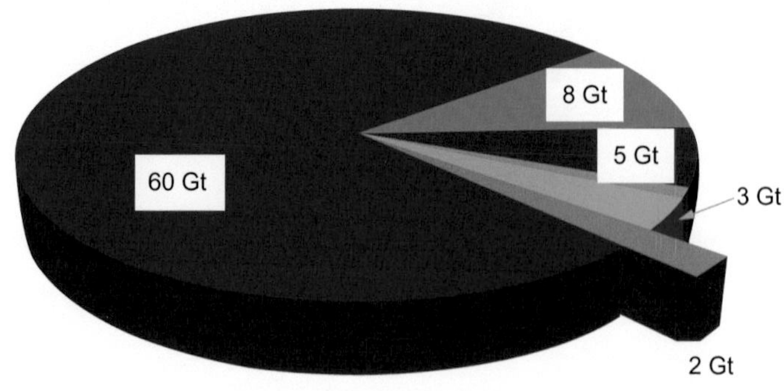

Fig. 5.9 British offshore storage space for CO_2. *(From presentation by David Clarke, Energy Technologies Institute, 2012. The evidence for deploying bioenergy with CCS (BECCS) in the UK. http://www.eti.co.uk/insights/the-evidence-for-deploying-bioenergy-with-ccs-beccs-in-the-uk.)*

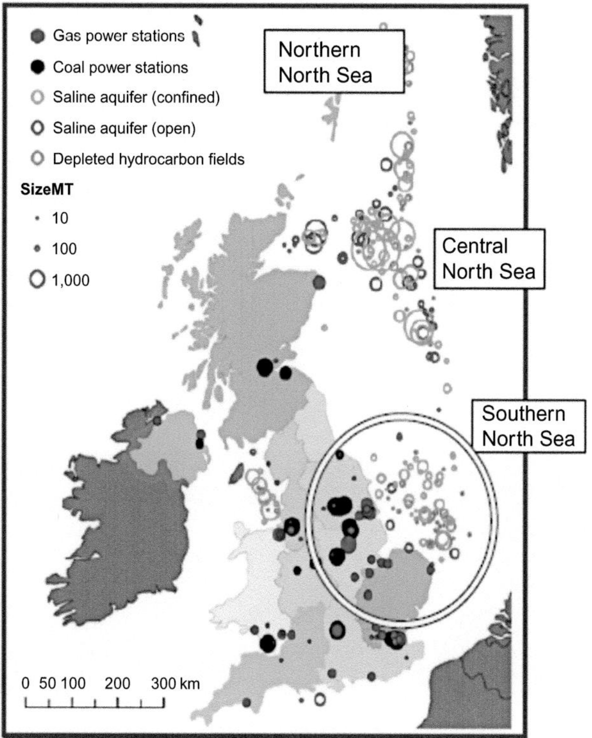

Fig. 5.10 Location of British offshore storage space for CO_2. *(From presentation by David Clarke, Energy Technologies Institute, 2012.)*

the rock. This means that simulation of CO_2 behaviour over very long periods is more complicated. What it does suggest, however, is that the longer the CO_2 is in the rock, the more securely it is disposed of because both dissolution and reaction will tend to form fluids or solids that are less likely to allow CO_2 escape.

Again, the rates of reaction can be simulated in computer models or scaled-down physical models. One of the latter is shown in Fig. 5.11.

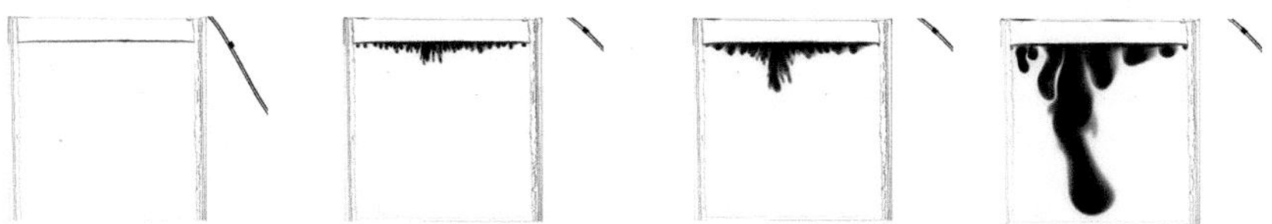

Fig. 5.11 Photographs of an experiment that shows CO_2 dissolution. The first square shows the experiment after 1 min 45 s; the second after 4 min 30 s; the third 7 min 30 s; and the last 30 min. *(Courtesy Chris Rochelle. BGS copyright NERC.)*

The diagram shows four tanks where the dissolution of CO_2 into saline water is demonstrated. The saline water has been treated to go brown as CO_2 dissolves. The tank is topped up with CO_2 gas. The first photograph shows the tank after 1 min 45 s; the second after 4 min 30 s; the third 7 min 30 s; and the last 30 min. CO_2 clearly dissolves in saline water quite quickly and begins to sink because it is denser than the original saline water. This illustrates an important principle in geological disposal—that at least one major process operating in the reservoir will tend to cause CO_2 to move downward, thus making it less able to escape. Computer simulations (Fig. 5.12) show similar patterns of dissolution.

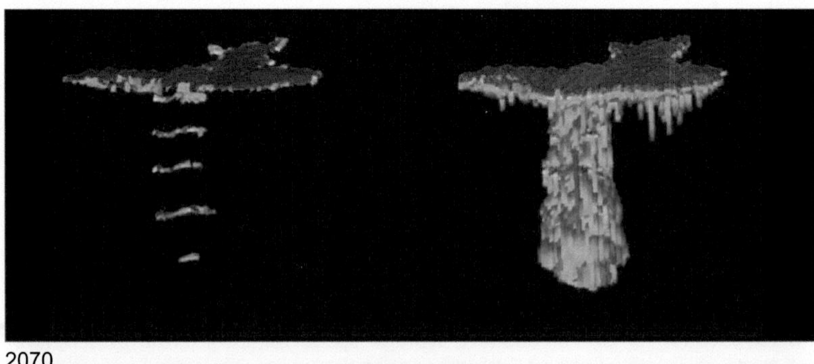

2070

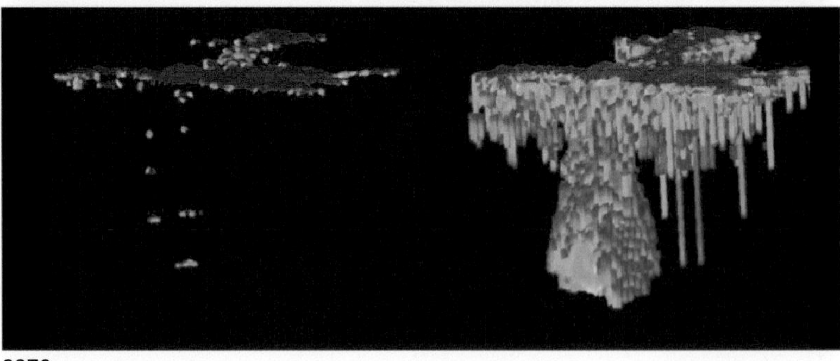

2270

Fig. 5.12 Simulations of the patterns of CO_2 dissolution at the Sleipner field in the North Sea. The simulations show the predicted state of the plume of CO_2 in Sleipner for 2070 and 2270. The left side of each image shows 'undissolved' CO_2, and the right shows CO_2 in solution. The colours represent the amounts of CO_2: red is high concentration, blue is low. *(From Chadwick, A. et al., 2008. Best practice for the storage of CO_2 in saline aquifers—observations and guidelines from the SACS and CO_2STORE projects. Nottingham, UK, British Geological Survey, 267pp. British Geological Survey Occasional Publication, 14.)*

Disposal in Depleted Oil and Gas Fields

The simplest solution to disposing of CO_2 is to inject into reservoirs that once contained oil and gas, sometimes called depleted oil and gas fields. These have advantages: they are unlikely to leak because they once confined oil and gas successfully, and as geological structures they are very well understood because they will have been studied and managed for many years as their load of oil and gas was extracted. Effectively the reservoir in a depleted oil and gas field can be regarded as like a tank or container to be used to store fluids, the most common being natural gas. Britain's largest natural gas storage facility—the Rough Field—was designed to store up to 2.8 billion m^3 of natural gas following its depletion (of its original gas) in the mid-1980s.

Depleted oil and gas fields are also convenient in the sense that they are serviced with wells, platforms, and pipelines that could be converted for use with CO_2.

The Sleipner Experiment

It is worth a small diversion to study the world's longest running industrial-scale CO_2 disposal project at Sleipner in the central North Sea. Here CO_2 is being injected into a thick layer of sandstone known as the Utsira Sand. The drilling platform stands in 80 m of water and the CO_2 injection point is at a depth of about 1012 m below sea level. The seismic section in Fig. 5.13 shows the Utsira Sand in orange and the main caprock in green.

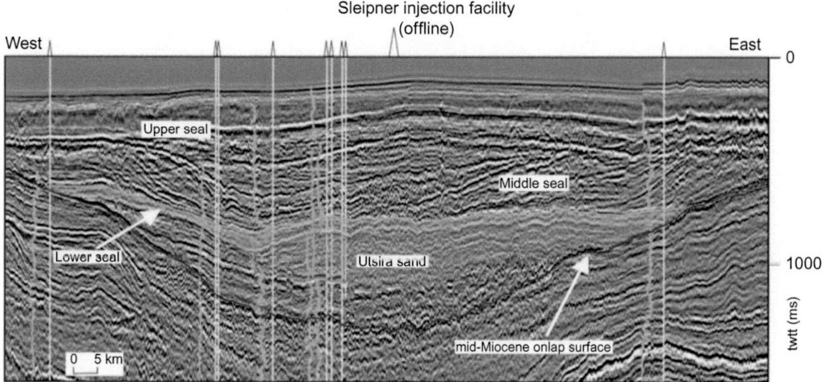

Fig. 5.13 The rock layers below the seabed near Sleipner. *(From Chadwick, A. et al., 2008. Best practice for the storage of CO_2 in saline aquifers—observations and guidelines from the SACS and CO2STORE projects. Nottingham, UK, British Geological Survey, 267pp. British Geological Survey Occasional Publication, 14. BGS copyright NERC. Reproduced with permission of the C02STORE Consortium.)*

The Utsira Sand extends widely under the North Sea and it varies from a few metres thick to about 300 m thick; thus it has an enormous volume. This coupled with its high porosity and permeability means that it is likely to have a very large capacity for CO_2.

Some advanced seismic analysis has enabled not just rock layers but actual CO_2 to be imaged on seismic profiles at Sleipner. Repeated seismic surveys have allowed images that show how CO_2 has collected since the start of injection in 1996 (Fig. 5.14).

This kind of detailed seismic to identify accumulated CO_2 can also be extended to the rocks above the reservoir to look for changes that might indicate leakage. The three images in the diagram (Fig. 5.15) show layers

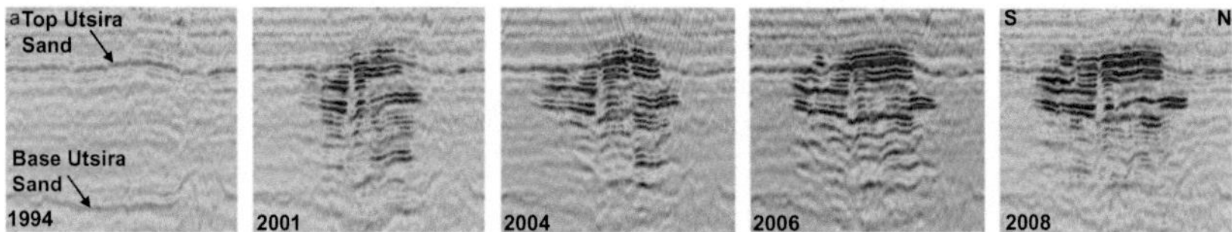

Fig. 5.14 The growth of the CO_2 plume at Sleipner. *(Courtesy Andy Chadwick. BGS copyright NERC.)*

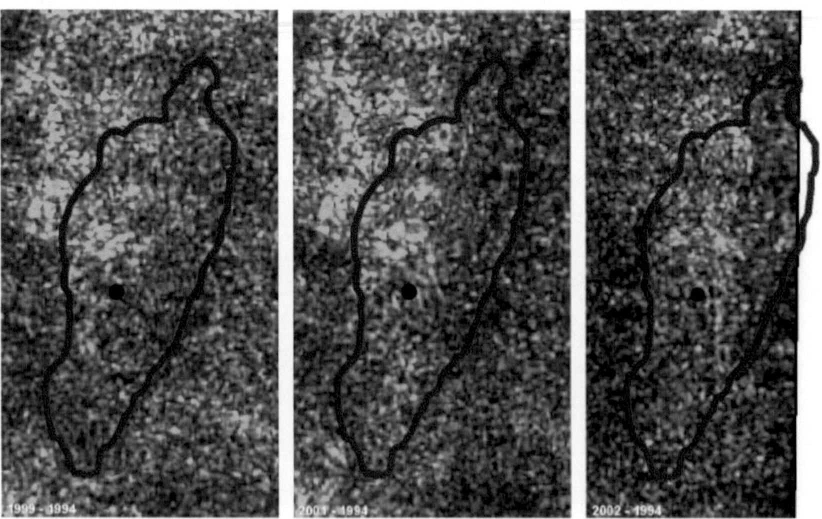

Fig. 5.15 The difference between the earliest seismic reflection data and later data for a level within the seal above the Utsira Sand. *(From Chadwick, A. et al., 2008. Best practice for the storage of CO_2 in saline aquifers—observations and guidelines from the SACS and CO2STORE projects. Nottingham, UK, British Geological Survey, 267pp. British Geological Survey Occasional Publication, 14. BGS copyright NERC. Reproduced with permission of the C02STORE Consortium.)*

above the reservoir (in plan view) with the area of CO_2 accumulation (traced from the reservoir below) outlined in red. The patterns following repeated surveys several years apart are the same, indicating no evidence of leakage, at least at the resolution or scale of a seismic survey.

How Will CCS Be Financed?

CCS will obviously be expensive because it involves at least three complex industrial activities: capture, transport, and storage (or disposal), none of which involve a revenue stream (unless CO_2 can be sold for enhanced oil recovery). A carbon tax or emissions trading scheme might be used to finance CCS.

An example is the European Union Emissions Trading Scheme (ETS) which is based on a 'cap and trade' principle. The 'cap', set by a government, is a limit on the amount of CO_2 a factory or a power station can emit. The companies are given permits to emit CO_2 up to the limit, but beyond that they have to buy a further permit (or permits) to emit more. The cost of buying these permits is an incentive to reduce emissions, if the price is high enough. If a power station is able to reduce its emissions below its limit, then it can also sell its unused permits to other emitters; this is known as 'carbon trading'. The system could be a cost-effective way of reducing emissions without more interventionist government measures. Presently the ETS enforces emissions reductions in 11,000 factories and power stations in 31 countries, but the price of EU ETS permits has been lower than expected, mainly because there is a surplus and so the incentive to reduce emissions has not been high.

Studies suggest that the first CCS projects in the power sector will cost between €60 and €90 per tonne of carbon dioxide captured and stored. But these costs are likely to come down as technology improves. Fig. 5.16 shows the range of the cost for carbon emission (carbon price forecast) and the decline in cost of CCS. The diagram shows that currently CCS is 'not economic on a standalone basis'. There may later be a commercial phase where the cost of CCS is similar or less than the carbon price, in which case CCS as a business would be viable and profitable.

BIOENERGY WITH CARBON CAPTURE AND STORAGE

Bioenergy with Carbon Capture and Storage (BECCS) is a variant on CCS that uses biofuels rather than fossil fuels as the source combustion material. The choice of combustion material is crucial because it improves the balance of energy and emissions such that BECCS could result in 'negative

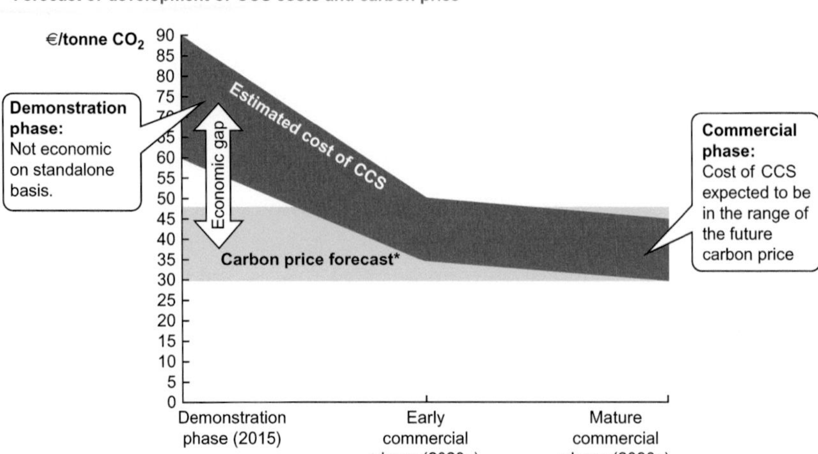

Fig. 5.16 Range of the cost for carbon emissions (carbon price forecast) and the decline in costs of CCS. *(From Carbon Capture and Storage: Assessing the Economics with permission from Reuters.)*

emissions'; in other words, the activity would result in a net extraction of CO_2 from the atmosphere. The different concepts are illustrated in Fig. 5.17. In conventional power production, fossil fuel from the subsurface is burned to produce electricity and then the CO_2 is released into the atmosphere. In so-called FECCS (fossil energy with CCS), the fossil fuel is burned and the resulting CO_2 is disposed of underground. This results in the CO_2 from that ancient geological environment essentially being returned to a geological environment. In a power station that is capable of capturing and disposing of all the CO_2 associated with its generating activity, this would mean zero emissions (ignoring the emissions associated with the building of the power station and so on). In a simple biofuel power station, the CO_2 emitted would be CO_2 that had, in the lifetime of the power station's fuel, been absorbed and used by the plant in photosynthesis. The idea then is that only CO_2 that was taken in by plants is being released in a cycle confined to the Earth's surface. Ignoring again the building of the power station and any emissions associated with the transport of biofuel to the power station, this suggests near-zero emissions.

Completing this theme is BECCS. Here, parts of the last two concepts are combined so that biofuel is burned and the resulting CO_2 is disposed of permanently in the subsurface.

Many of the Intergovernmental Panel on Climate Change (IPCC) scenarios for emissions reduction include some form of 'negative emissions' or

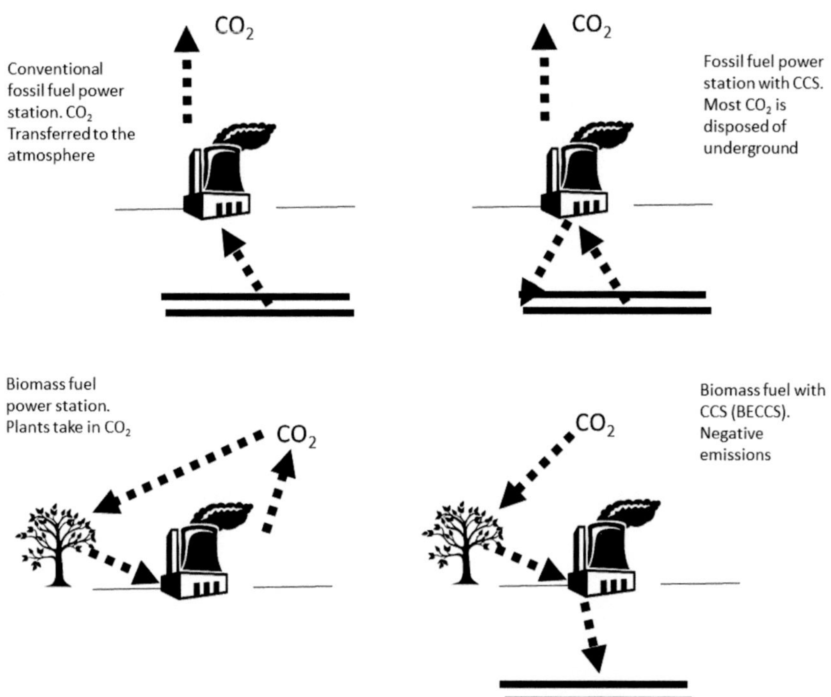

Fig. 5.17 BECCS is a variant on CCS that uses biofuels rather than fossil fuels as the source combustion material. The choice of combustion material is crucial because it improves the balance of energy and emissions such that BECCS could result in 'negative emissions'.

permanent net removal of GHG emissions from the atmosphere. Of the 400 IPCC climate scenarios that have a 50% or better chance of less than 2°C warming, more than 300 assume the successful and large-scale uptake of negative-emission technologies.

So BECCS is being taken seriously. Among the advocates is the UK's Energy Technology Institute (ETI), which believes that BECCS should be part of British energy strategy. Noting that Britain already uses biomass extensively in power generation, ETI estimates that for an effective level of BECCS, additional domestic biomass feedstock production needs could be met by converting 1.4 million ha of UK agricultural land to bioenergy crops and forestry by the 2050s, and that there is sufficient spare land in the UK agricultural system to meet this land requirement, without impacting existing levels of UK food security.

ETI considered crops such as *Miscanthus* (silvergrass), 'short rotation coppice willow' and 'short rotation forestry' grown on arable land or grassland in their modelling. These are high-yield crops, but also could deliver wider biodiversity and ecosystem service benefits, including flood mitigation, pest control, and wildlife habitat.

ETI suggested that the negative emissions of BECCS provide 'head-room' or 'breathing space' because they reduce the need for rapid emissions reductions in sectors such as heavy-duty transport and aviation, which are more difficult and expensive to decarbonise, at least with present technology. ETI's modelling suggests BECCS could deliver about 55 million tonnes of net negative emissions per annum (approximately half the UK's emissions target in 2050), whilst also meeting about 10% of the UK's future energy demand.

Some progress has been made in demonstrating BECCS in that the multinational Toshiba is aiming to capture over 500 tonnes of CO_2 per day from the Mikawa Power Plant in Japan, which by 2020 will also be able to fire biomass as well as coal. Thus, Mikawa will become the world's first power plant capable of capturing carbon following biomass combustion.

In a recent article in the magazine *Science*, Johan Rockström and colleagues proposed a global carbon law to halve gross anthropogenic CO_2 emissions every decade. According to Rockström, by the 2020s there must be carbon pricing across the world to cover all GHG emissions. By 2030 all building construction must be carbon-neutral or carbon-negative and BECCS schemes totalling 1–2 $GtCO_2$/year must be in operation (Fig. 5.18).

Predictably with such a controversial idea, other views are much less favourable toward BECCS. The arguments against it range from concerns that BECCS's negative emissions lessen the imperative to act quickly and may allow governments and agencies to become complacent. This is known as the 'pay later' scenario where BECCS provides the possibility of a slower phase-out of high carbon technology. But of course this is only if BECCS delivers negative emissions; if it fails, then the 'pay later' plan is dangerously flawed.

By far the most important questions over the feasibility of BECCS are how much land and resources can be devoted to biofuel crops, and the amount of subsurface CO_2 disposal space. The first is a difficult problem. In a world where population is growing and land and other resources are at a premium, can space be devoted to crops that we simply burn? Many think not. A paper by Lydia Smith and Margaret Torn in the journal *Climatic Change* is sceptical of the very high sequestration potentials for BECCS that have been reported, pointing out that there has been no systematic analysis

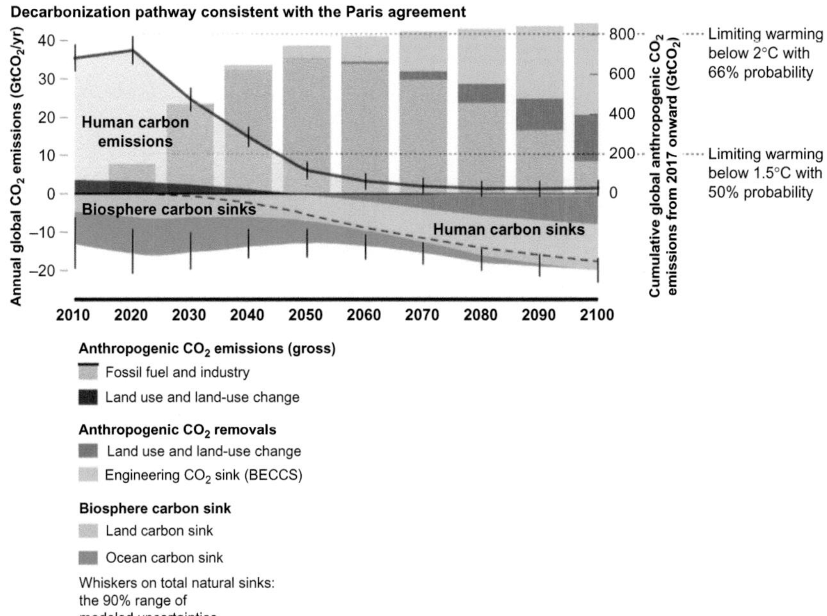

Fig. 5.18 Decarbonisation pathway consistent with the Paris Agreement, including BECCS. *(From Rockström, J. et al., 2017. A roadmap for rapid decarbonisation. Science 355, 1269–1271.)*

of the potential ecological limits to, and environmental impacts of, implementation at the scale relevant to *global* climate change mitigation. In a simple model they estimated that to remove 1 PgC/y (1 petagram carbon per year—modern fossil fuel use emits about 8 PgC/y) by burning biofuel would require at least 2×10^8 ha of land (20 times US area currently under bioethanol production) and 20 Tg/y (teragrams per year) of nitrogen (20% of global fertilizer nitrogen production), consuming 4×10^{12} m^3 per year of water. Smith and Torn also considered the lifecycle emissions and losses of CO_2 that might be associated with biomass production and processing with carbon capture and storage, concluding that efficiency would probably be low in that from crop to geological reservoir a net sequestration of 1 PgC would require the fixing of 2.1 PgC (Fig. 5.19).

Is there enough geological storage space in the parts of the world that produce significant biofuels to avoid expensive and carbon-consuming transport of captured CO_2? Little is known about this. In the past CCS has focused on coal, and capture and storage has been linked to conventional

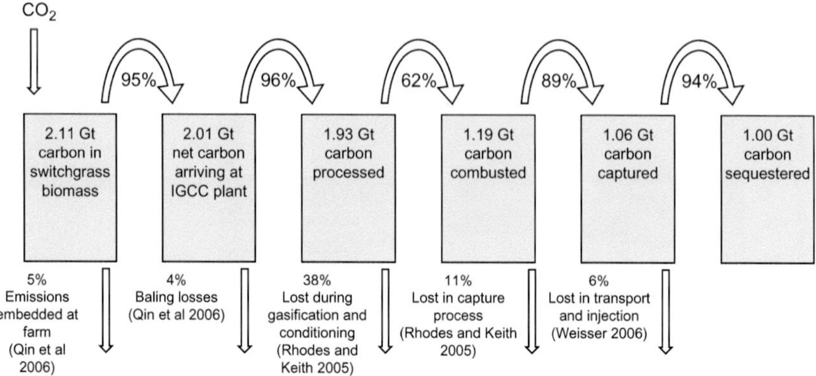

Fig. 5.19 Carbon flow, life cycle emissions, and carbon losses during temperate switchgrass production and processing with carbon capture and storage. *(For details see Smith, L.J., Torn, M.S., 2013. Ecological limits to terrestrial biological carbon dioxide removal. Climatic Change 118, 89–103.)*

industrial areas. Would agricultural sources for power station feedstock alter the geographical pattern of power supply and generation?

These uncertainties are rightly pointed out by critics of BECCS. They see the politics of climate change running ahead of the science. Not only do they point out that BECCS (and CCS for that matter) are unproven at an industrial scale, but also that BECCS could encourage fossil fuel lock-in and perhaps a lazy attitude to really tackling the problem, as in the 'pay later' scenario.

INDIRECT WAYS THAT GEOLOGICAL MATERIALS CONTRIBUTE TO LOW CARBON

The first part of this chapter examined ways that geological intervention within the carbon cycle could directly reduce emissions through CCS and BECCS. However, geological materials also provide less direct contributions to low carbon energy.

Geothermal provides access to heat for electricity production and heating of buildings and thus offers very low carbon energy. Three other areas deserve mention as more indirect ways that geological materials and the geosphere influence how well we adapt to the climate/energy challenge. These have a principle in common—the principle that rocks provide secure storage or disposal potential. They include compressed air energy storage (CAES), hydrogen storage, and radioactive waste disposal. These are all proven technologies that are in use, though not to their full potential. They do not directly involve emissions control, but facilitate other technologies or groups of activities that fit into a low-carbon energy system.

Geothermal Energy

Perhaps the most obvious way that the Earth's materials can contribute to low-carbon energy is through provision of geothermal energy.

The Earth's indigenous natural heat is partly residual in that it is left over after the formation of the Earth, and partly self-generated in that radioactive decay in the deep subsurface generates heat. There is more heat at great depth than close to the surface; for example, temperatures at the core–mantle boundary may reach over 4000°C, and the temperature decreases in a geothermal gradient up through the crust. High heat flow at great depth causes rocks to behave plastically, so that in the Earth's mantle large convection currents exist, which are ultimately responsible for plate movements and for most earthquakes and volcanoes.

Calculations suggest that the total heat present in the Earth is around 12.6×10^{24} megajoules (MJ), and the proportion in the crust is around 5.4×10^{21} MJ. These figures dwarf the size of present world electricity generation, which is less than 10×10^{13} MJ. But as I showed in an earlier chapter, this could be regarded as a resource rather than a reserve. The thermal energy of the Earth is huge but most of this is well out of our reach, so that only a tiny amount could ever be used. In many cases the reason why we can use geothermal energy is because it is brought from deeper in the Earth by a 'carrier', usually water.

Broadly, geothermal can be divided into two types. Heat that is sufficient to generate electricity, and heat that is sufficient only for supplementing heating systems in buildings or for industrial processes. By 2017 about 13.4 GW of geothermal electricity was being produced from power stations; but a much larger amount of power, about 28 GW, is provided for direct heating of houses and public buildings, spas, industrial processes, desalination and glass houses. Advocates of geothermal energy believe that exploiting present technology in a wider range of sites would make 70 GW of geothermal electricity possible across the world—and that new technologies could increase that figure to 140 GW.

Conventional electric power production is commonly limited to fluid temperatures above 180°C, but with binary fluid technology lower temperatures can also be used to generate electricity down to about 70°C. For direct district heating, useful temperatures range from 80°C to just a few degrees above the ambient temperature.

At least 90 countries have potential geothermal resources, though only about 70 tap this potential. Electricity is produced from geothermal energy in only 24 countries. Interestingly, developing countries, for example in East Africa, are active in both geothermal electricity and direct heat. China tops the list of countries using direct geothermal for municipal heating. As discussed earlier, geothermal alongside other renewables could help developing countries avoid a large-scale fossil economy.

The UK has often been considered to have a rather modest geothermal resource; however, new research led by Jon Busby has revealed that by developing more efficient methods of extracting heat, even a relatively tepid underground like that of the UK could yield significant heat. The key extraction method is known as the engineered or enhanced geothermal system (EGS). An EGS differs from conventional geothermal extraction in that rocks containing heat or warm fluids are artificially fractured by high-pressure water. This opens cracks in the warm rocks and allows introduced water or indigenous fluids to flow more easily as carriers of the heat. Assuming application of EGS, the total heat in place in the deep rocks of the UK (to a depth of 9.5 km) is about 360,000 exajoules (an exajoule (EJ) is 10^{18} joules). Busby points out that '…if it were possible to develop just 2% of this resource, this would be equivalent to 124 times the final UK energy consumption in 2015.'

As one might expect, the carbon footprint of most kinds of geothermal energy is very low because heat is extracted with little energy expended, though the carbon intensity of infrastructure and for EGS (for energy related to hydraulic fracturing) would have to be taken into account.

Compressed Air Energy Storage

Compressed air has long been used to store energy; for example, pressurised air tanks are used to start diesel generators and to propel underground mine railways. In geological CAES, the idea is to store large amounts of compressed air in underground caverns—mainly in salt layers—for extraction through a turbine later.

The challenge of intermittency raises the problem of being able to store excess energy when it is fed into the grid on windy and sunny days, and

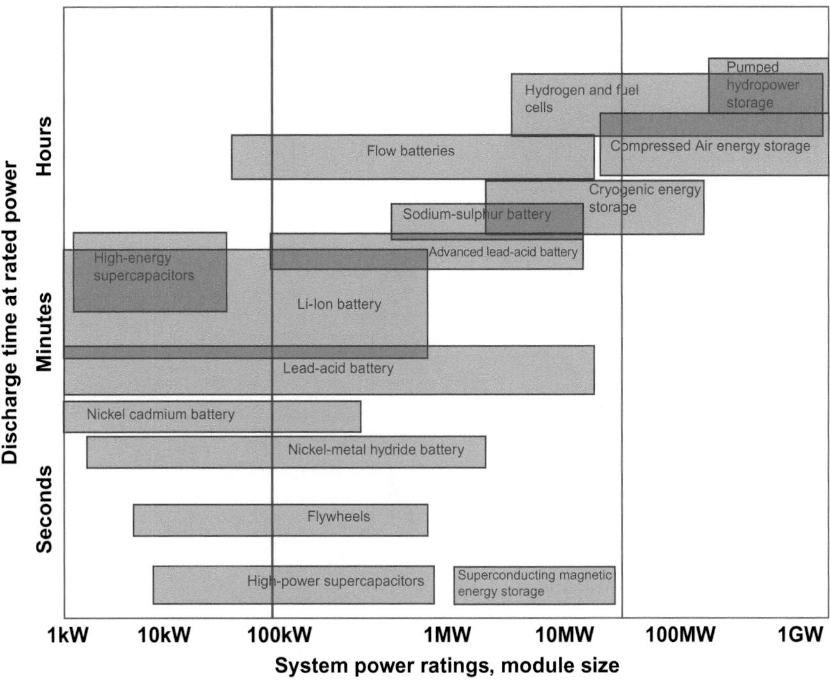

Fig. 5.20 The suitability of different energy storage technologies for grid-scale applications. *(After the Centre for Low Carbon Futures website: http://www.lowcarbonfutures.org/pathways-energy-storage-uk.)*

being able to use it later when demand rises above supply. The challenge can be reduced to two variables: the capacity of storage and the speed at which the stored energy can be made available to the grid. This is conceptualised in Fig. 5.20. A few technologies provide grid-scale capacity (or 'bulk power management'), including pumped hydroelectric storage and CAES, even though the speed at which the power can be accessed is slower than, for example, from large batteries.

One of the engineering challenges for CAES, which anyone who has ever used a bicycle pump will know, is that air heats up when compressed from atmospheric pressure and in an industrial CAES situation a storage pressure of about 70 bar is envisaged. Heat must be controlled to avoid damage to compressors and caverns.

Salt caverns are favoured because, being impermeable, there are no pressure losses, and because there is no reaction between the oxygen in the air and the salt. The UK has relatively large onshore areas underlain by salt, some of which are already used for natural gas storage (Fig. 5.21).

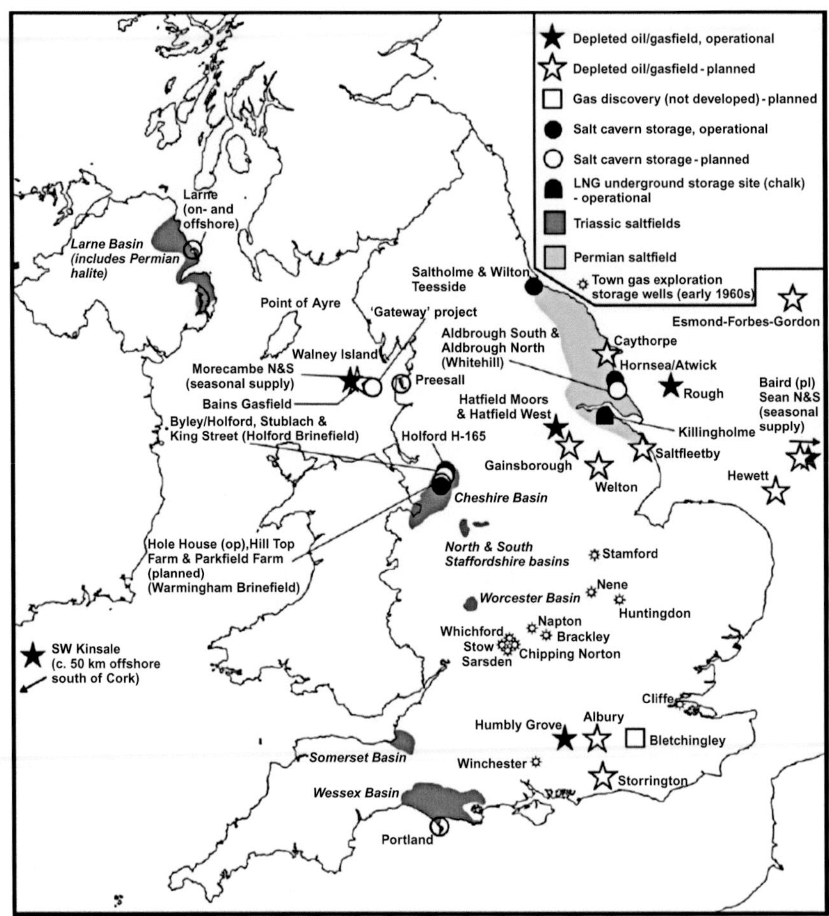

Fig. 5.21 Location of operational and proposed UK underground gas storage sites, including depleting oil and gas fields and mined chalk facilities. Also shown are the locations of the main salt-bearing basins onshore. *(From Evans, D. et al., 2009. The use of Britain's subsurface. Land Use Policy 26 (1), S302–S316.)*

CAES is possible in natural aquifers, though oxygen may react with minerals in the host rock and microorganisms in an aquifer can deplete oxygen and alter the character of the stored air; similarly, bacteria can act to block pore spaces in the reservoir. Depleted natural gas fields could also be used for CAES, though any mixing of residual hydrocarbons with compressed air would have to be considered.

CAES has advantages over grid-scale batteries, including longer lifetimes of pressure vessels and lower material toxicity. However, cavern design and construction are expensive.

Earth Resources for Batteries for Transport and Grid-Scale Storage

Like CAES, batteries may provide grid-scale energy storage, and also in the future may power more and more of the world's private and commercial transport. The overall 'lifetime costs' of owning and driving an electric car may soon be close to those for cars with an internal combustion engine, so that by 2030 perhaps most private cars will be electric. Forecasters such as the media company Bloomberg suggest that there will be 500 million electric cars by 2040.

Battery technology may also be part of the solution for grid-scale electricity. Batteries are already used to even out power distribution and are often colocated with renewable energy plants. In small power distribution networks for remote areas or islands, batteries may provide stabilisation, as they already do in Puerto Rico and in parts of Alaska.

Batteries may seem far away from considerations of geological science, but in fact an understanding of the primary resources used to make batteries is crucial. These include deposits of lithium, sodium, vanadium, copper, cobalt, and nickel.

A recent estimate from a spokesman from the mining company Glencore, based on electric vehicles being 30% of the global vehicle fleet by 2030, suggested that an extra 2 million tonnes of copper, 1.2 million tonnes of nickel and 260,000 tonnes of cobalt will have to be mined per year into the future. These are considerable increases on present production levels and suggest that more resources will have to be found, and that recycling of materials will have to be improved.

Hydrogen

The term 'hydrogen economy' was first coined by the physicist John Bockris as an alternative to the present hydrocarbon economy. The hydrogen economy encompasses fuel for transport (road vehicles and shipping), stationary power generation (for heating and power in buildings), and an energy storage medium feeding from off-peak excess electricity. A large-scale change from hydrocarbons to hydrogen as the 'prime mover' requires a radical rethink of infrastructure and storage, for example geological storage and pipelines for hydrogen, not unlike the infrastructure presently in place for natural gas and being discussed for CO_2 and CCS.

The storage technology of compressed hydrogen gas storage in salt caverns is similar to that of natural gas; however, hydrogen energy density by

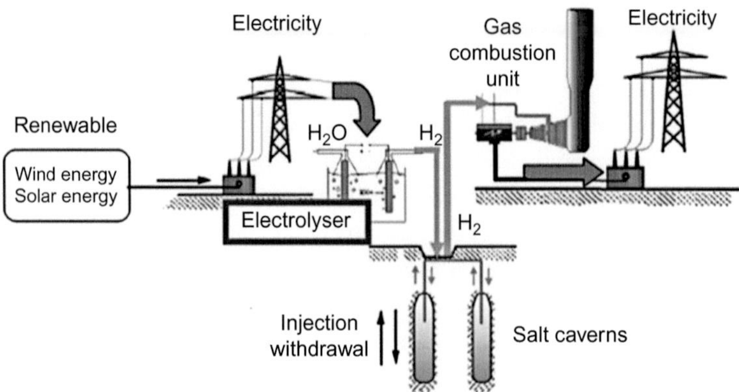

Fig. 5.22 A system to produce hydrogen, store hydrogen, and produce electricity from hydrogen. *(From Ozarslan, A., 2012. Large-scale hydrogen energy storage in salt caverns. Int. J. Hydrogen Energy 37, 14265–14277.)*

volume is only one-third of that of natural gas and so gaseous hydrogen energy storage is more expensive. To maximise efficiency, hydrogen gas is compressed in underground salt caverns to a pressure of 20 MPa (200 bar) or more.

A system for hydrogen generation, salt cavern storage, and electricity generation is shown in Fig. 5.22. Wind and solar energy generate renewable electricity. At times of excessive electricity production or other times, electrolysers can use the electricity to produce hydrogen and oxygen from water. The hydrogen is stored below the plant in a salt cavern. A gas combustion power plant using hydrogen alone or combined with natural gas can generate electricity. Excess renewable electricity can also be used to produce hydrogen from natural gas.

Hydrogen storage is not new; hydrogen is already stored in a small number of salt caverns in the UK and the United States, supporting chemical plants and oil refineries. The largest single store in the United States holds over 100 GWh (energy equivalent) of hydrogen. The extremely low permeability of salt makes it the best candidate for storing hydrogen's small molecule.

In this book I have examined energy transitions—from wood to coal, and from coal to hydrocarbons. What might a transition to hydrogen look like? The systems analyst Leonardo Barreto envisages a relatively smooth transition toward a post-fossil global energy system, illustrated cleverly in a ternary diagram where oil/gas, coal and zero-carbon are the apices (Fig. 5.23). Barreto sees fossil fuels dominating the primary energy supply

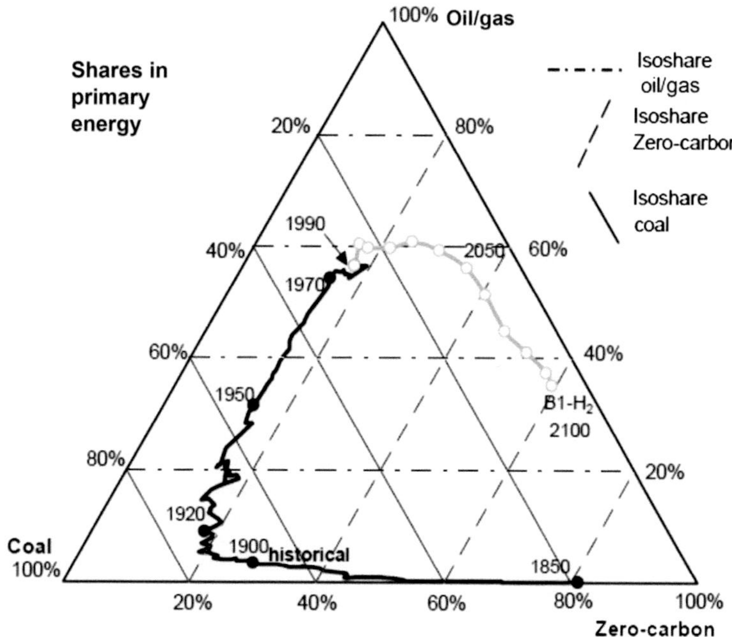

Fig. 5.23 Global shares in primary energy use, coal, oil/gas, and non-fossil energy, illustrated with an "energy triangle" (in percent). *(From Barreto, L. et al., 2003. The hydrogen economy in the 21st century: a sustainable development scenario. Int. J. Hydrogen Energy 28, 267–284.)*

until 2050, but during this period the system shifts toward natural gas, which operates as the main transitional fuel to the post-fossil era. Hydrogen becomes the main 'energy carrier' by the end of the 21st century.

Interestingly, hydrogen, as well as being capable of being manufactured, also occurs naturally as a gas in rock reservoirs, rather like hydrocarbons. Geological hydrogen mainly forms in the subsurface due to chemical alteration of ultramafic igneous rocks and associated serpentinite. Huge volumes of ultramafic igneous rock are present in the Earth's mantle and are therefore inaccessible, but masses of ultramafic rocks are sometimes associated with fragments of deep crust that have been detached during unsuccessful tectonic subduction. Such masses, known as ophiolites, also come into contact with groundwater, leading to serpentinisation, a process of hydrolysis of ferromagnesian minerals which produces hydrogen gas and the mineral serpentine [$(MgFe)_3Si_2O_5\ (OH)_4$].

There are no estimates for the resource of geological hydrogen worldwide, though the Semail ophiolite in southern Arabia, the largest in the world, could have produced 3125 km^3 of hydrogen. To date there is no commercial production of geological hydrogen.

Geological Radioactive Waste Disposal

Nuclear energy is widely considered to be a contributor to low-carbon power production and nuclear power plants the world over have produced useful power, but also radioactive waste. The UK, for example, has accumulated a substantial legacy of radioactive waste since the 1940s and will continue to do so for many years into the future. By 2100, it is likely that 2.6 million tonnes of high-level radioactive waste will need to be safely managed, probably within deep caverns constructed specifically for the purpose. Essentially, a geological disposal facility (GDF) makes use of engineered materials and structures, including concrete, metals and clays, as well as the surrounding geological environment, as containment barriers (Fig. 5.24).

A big part of containment is the natural arrangement of the rocks that surround the engineered barriers. In many ways this is no different from underground disposal or containment of CO_2, for example; however, radionuclides may be hazardous for up to a million years into the future. Thus a fundamental requirement of the geological environment is that its behaviour should be predictable enough to establish very long-term radiological safety. Amongst the factors that need to be assessed are present and future seismic activity, glaciation, uplift and erosion, climate change including sea-level

Fig. 5.24 A geological disposal facility (GDF). *(From https://www.gov.uk/government/organisations/radioactive-waste-management.)*

rise, isostasy, and permafrost formation—because all of these processes could compromise the GDF. An assessment of risk involves rather detailed study of geological processes occurring now and in the recent past in order to understand changes up to 1 million years into the future—and is a geological form of futurology.

An example is the seismic hazard threat to a GDF. One way to estimate this is to look at the spatial distribution of previous earthquakes in a given region, the magnitude and recurrence of those earthquakes, and the likely ground motion (or 'ground acceleration') that could result at different distances from the origin of the earthquake energy. A map can be computed showing statistical likelihood of ground acceleration (Fig. 5.25).

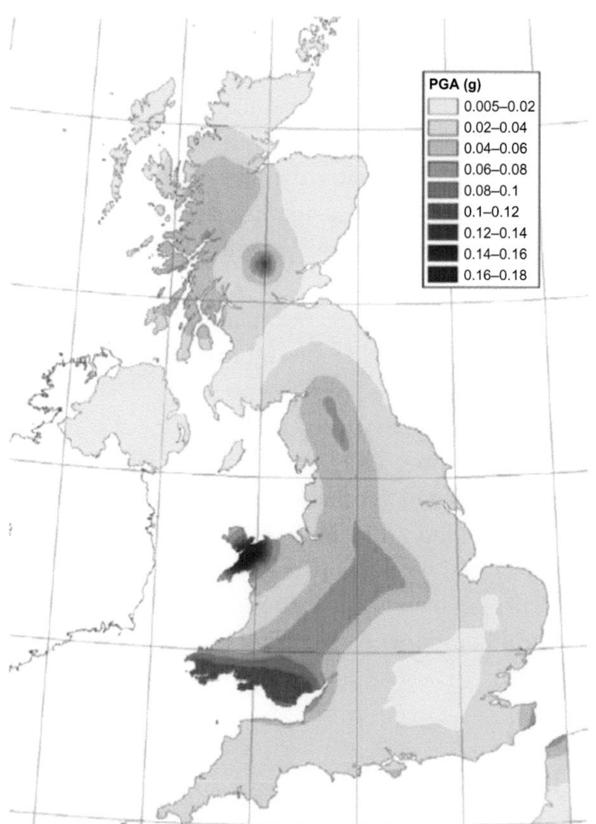

Fig. 5.25 Seismic hazard with reference to a GDF. The map shows peak ground accelerations (PGA) with a 10% probability of being exceeded for a 2500-year return period. *(From McEvoy, F.M. et al., 2016. Tectonic and climatic considerations for deep geological disposal of radioactive waste: A UK perspective. Sci. Total Environ. 571, 507–521.)*

A part of the futurology that I mentioned has to look into the form of warning that might be placed at the surface above a GDF, which could be recognised by Earth's inhabitants many thousands of years into the future. Geologists, linguists, and semiotics experts that work on this agree that whatever is placed at the surface above a GDF must have a clear message that will survive and be visible—and be comprehensible and trusted. It may be that GDFs will be marked by pyramid-like earth structures and permanent symbols and messages in many languages that convey the appropriate warning.

SUMMARY

The geological parts of CCS—injection, storage capacity, reservoir management, and the long-term fate of CO_2 in rocks—are increasingly well understood. Though great strides are being made in renewables decarbonisation of power production, the chances are that at least some industrial-scale CCS will be needed to deal with emissions from industry, for example cement factories and refineries. The greater the take-up of renewables, the greater the need for intermittency to be accommodated for by load-following gas power stations. Whether these are abated with CCS depends on the permissible emissions. Rather worryingly, BECCS as a net negative emissions technology is seen by some as the last resort if emissions are not cut quickly enough. If the developing world takes up fossil fuels, perhaps BECCS will become important.

The wide range of ways that deep rocks might act as sites for storage or disposal, for example for CO_2, compressed air, nuclear waste and hydrogen—or for sources of geothermal energy, underlines the importance of geological studies of fluid flow in rocks and how the subsurface, like any other part of the natural world, is managed into the future.

Geologists will also have to give thought to the Earth's resources that provide the critical components of batteries, such as lithium and sodium.

BIBLIOGRAPHY

Anderson, K., 2015. Duality in climate science. Nature Geosci. 8, 898–900.
Barreto, L., et al., 2003. The hydrogen economy in the 21st century: a sustainable development scenario. Int. J. Hydrogen Energy 28, 267–284.
Busby, J., Terrington, R., 2017. Assessment of the resource base for engineered geothermal systems in Great Britain. Geotherm. Energy 5, 7.
Chadwick, A., et al., 2008. Best practice for the storage of CO_2 in saline aquifers—observations and guidelines from the SACS and CO_2STORE projects. British

Geological Survey, Nottingham, UK, 267pp. British Geological Survey Occasional Publication, 14

Dickson, M.H., Fanelli, M., 2003. Geothermal Energy: Utilization and Technology. UNESCO, Paris, p. 221.

Energy Technologies Institute, 2012. The evidence for deploying bioenergy with CCS (BECCS) in the UK. http://www.eti.co.uk/insights/the-evidence-for-deploying-bioenergy-with-ccs-beccs-in-the-uk.

Evans, D., et al., 2009. The use of Britain's subsurface. Land Use Policy 26 (1), S302–S316.

Fridleifsson, I.B., 2001. Geothermal energy for the benefit of the people. Renew. Sustain. Energy Rev. 5, 299–312.

Gammer, D. et al. 2011. The UKSAP Consortium. The Energy Technologies Institute's UK CO_2 Storage Appraisal Project (UKSAP). SPE Paper Number 148426.

Gibbins, J., Chalmers, H., 2008. Preparing for global rollout: A 'developed country first' demonstration programme for rapid CCS deployment. Energy Policy 36, 501–507.

Holloway, S., 2007. Carbon dioxide capture and geological storage. Philos. Trans. Royal Soc. Lond. A 365, 1095–1107.

Holloway, S., et al., 2005. A review of natural CO2 occurrences and releases and their relevance to CO2 storage, BGS External Report CR/05/104. 117pp.

Kemp, A.G., Kasim, S.A., 2010. A futuristic least-cost optimisation model of CO_2 transportation and storage in the UK/UK continental shelf. Energy Policy 38, 3652–3667.

Letcher, T., 2016. Storing Energy with Special Reference to Renewable Energy Sources 1st Edition. Elsevier. 590pp.

McEvoy, F.M., et al., 2016. Tectonic and climatic considerations for deep geological disposal of radioactive waste: A UK perspective. Sci. Total Environ. 571, 507–521.

Metz, B., et al., 2005. Carbon Dioxide Capture and Storage IPCC. Cambridge University Press, UK, p. 431.

Musson, R.M.W. and Sargeant, S.L., 2007. Eurocode 8 Seismic Hazard Zoning Maps for the UK. British Geological Survey, Technical Report CR/07/125N. Unpublished.

National Energy Technology Laboratory (NETL), 2010. Carbon Sequestration Atlas of the United States and Canada, third ed. 162pp.

Ozarslan, A., 2012. Large-scale hydrogen energy storage in salt caverns. Int. J. Hydrogen Energy 37, 14265–14277.

Pacala, S., Socolow, R., 2004. Stabilization wedges: solving the climate problem for the next 50 years with current technologies. Science 305, 968–972.

Rockström, J., et al., 2017. A roadmap for rapid decarbonisation. Science 355, 1269–1271.

Smith, L.J., Torn, M.S., 2013. Ecological limits to terrestrial biological carbon dioxide removal. Climatic Change 118, 89–103.

Smith, D.J., et al., 2011. The impact of boundary conditions on CO_2 storage capacity estimation in aquifers. Energy Procedia 4, 4828–4834.

Stephenson, M.H., 2013. Returning Carbon to Nature: Coal, Carbon Capture, and Storage. Elsevier, Amsterdam, Netherlands. 143pp.

Stephenson, M.H., 2014. Five unconventional fuels: geology and environment. In: Unconventional Fossil Fuels: The Next Hydrocarbon Revolution?. Emirates Center for Strategic Studies and Research, Abu Dhabi, pp. 13–34.

Stephenson, M.H., Rochelle, C., 2010. Have your coal and burn it. Educ. Chem. 47, 23–25.

Zheng, K., Zhang, Z., Zhu, H., Liu, S., 2005. Process and prospects of industrialized development of geothermal resources in China—Country update report for 2000–2004. In: Proceedings of the World Geothermal Congress 2005, Antalya, Turkey.

CHAPTER 6

Climate Change Adaptation: Geological Aspects

Contents

Modern climate change modelling indicates change across the world affecting temperature and precipitation; it also indicates that one of the most important 'carriers' of climate change will be river runoff and that declining runoff will result in greater water stress. In many areas, particularly in developing countries, groundwater has the potential to be very useful in providing backup for surface water as a buffer for climate change, though it is not clear how climate change will affect groundwater availability, nor as the developing world grows and industrialises where the most intense future needs will be. Research into the positions of present and future 'development corridors' may allow greater understanding of the geographical locations of pressure points.

Around the world, rocks also provide storage for excess water, for example in cities where underground reservoirs and engineered drains using natural material can help deal with flooding.

In earlier chapters I have tried to show how the subsurface of the Earth and the materials contained within are important to two of the big themes of the modern world: energy and climate change. In this section I will show how the subsurface will play a big part in how humankind adapts to climate change. The effects of climate change and adaptation will be approached differently in different parts of the world, but general global changes are easier

Energy and Climate Change
https://doi.org/10.1016/B978-0-12-812021-7.00006-3

to predict than local changes. Developed and developing countries will also have different challenges, vulnerabilities, and solutions to adaptation. In this book I look particularly at the United Kingdom as an example of a developed country, and a selection of developing countries.

MODELLING AND CLIMATE CHANGE

In the chapter on deep-time climate change, I showed that the long-term geological carbon cycle can have an enormous effect on climate, and that previous periods of rapid global warming can be shown to have had roots in a changing atmosphere where more greenhouse gases are present. Deep-time climatologists can show the big climatic and environmental trends in spatial and temporal terms. However, what is needed for decision makers are local detailed predictions and forecasts that allow planners to design (for example) flood barriers in a city, sea defences at the coast, or make sure that large building infrastructure is secure against changing climate. Uppermost in the planners' minds would be how high to make the sea defences or the flood barriers, how to 'future-proof' the infrastructure, and, ultimately, what is the most cost-effective use of public money to guard against the effects of climate change.

For this, planners and policymakers go to climate modellers that build on our understanding of present weather and relatively recent historical climate change. Essentially climate change forecasting is weather forecasting on a much longer timescale. The main tool is the general circulation model (GCM), based on mathematical representations of physical processes. GCMs use big computing power to match past climate data, link causes and effects in climate change, and make forecasts.

The most authoritative recent global forecast is the IPCC's Fifth Assessment Report, published in 2014. It contains a 'Summary for Policymakers' including general information with attached levels of certainty. This includes information on observed impacts of climate change so far, including on water resource quantity and quality, and glaciers and permafrost thaw. It also details the shift in geographic ranges and habits of terrestrial, freshwater, and marine species (Fig. 6.1).

The summary also points out eight main global risks from future climate change:

1. Risk of harm on low-lying coasts and small islands, due to storm surges, coastal flooding, and sea-level rise
2. Risk of harm for large urban populations due to inland flooding in some regions

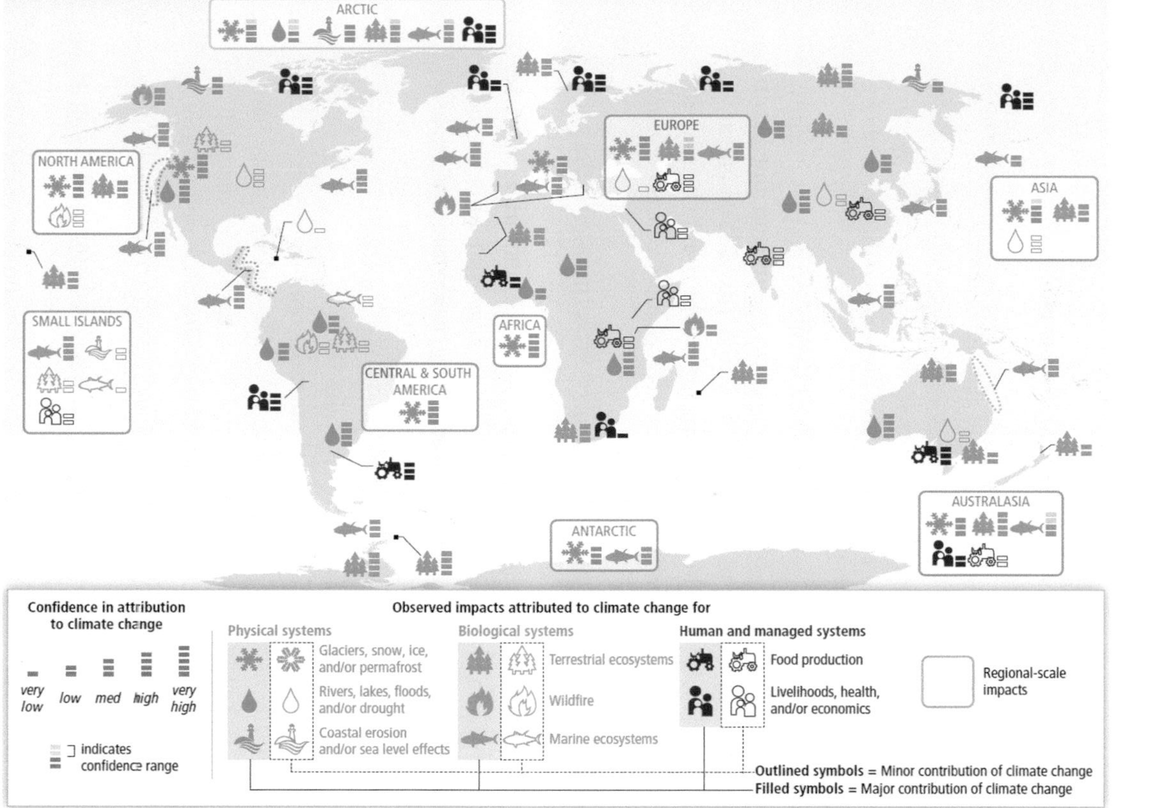

Fig. 6.1 Global patterns of impacts in recent decades attributed to climate change. *(Figure SPM.2. From IPCC, 2014. Summary for policymakers. In: Field, C.B., Barros, V.R., Dokken, D.J., Mach, K.J., Mastrandrea, M.D., Bilir, T.E., Chatterjee, M., Ebi, K.L., Estrada, Y.O., Genova, R.C., Girma, B., Kissel, E.S., Levy, A.N., MacCracken, S., Mastrandrea, P.R., White, L.L. (Eds.), Climate Change 2014: Impacts, Adaptation, and Vulnerability. Contribution of Working Group II to the Fifth Assessment Report of the Intergovernmental Panel on Climate Change. Cambridge University Press, Cambridge, United Kingdom and New York, NY.)*

3. Risk of harm due to extreme weather events leading to breakdown of critical services such as electricity, water supply, and health and emergency services
4. Risk of harm during periods of extreme heat, particularly for vulnerable urban populations and those working outdoors in urban or rural areas
5. Risk of food insecurity, particularly for poor people
6. Risk of loss of rural livelihoods and income due to scarce water, particularly for poor farmers and pastoralists in semiarid regions
7. Risk of loss of marine and coastal ecosystems and therefore coastal livelihoods, especially in fishing communities in the tropics and the Arctic
8. Risk of loss of inland water ecosystems and the services they provide for livelihoods

Though this information is clearly of use to policymakers in that it allows general planning at the level of governmental environmental and health budgets, it is not suitable for planners trying to understand local effects. It is, after all, at the local level where climate change will be mostly keenly felt. In this section I will look in more detail at the UK and developing countries in particular.

In the UK, the Government 'UK Climate Projections' are the most authoritative and comprehensive source of information. Like the IPCC, the UK Climate Projections (the latest are from 2009 and are known as UKCP09) provide information on observed changes and forecasts but, focusing in on the local, they also offer marine and coastal forecasts, including future projections for sea-level rise, storm surge, sea temperature, salinity, current, and waves. The UKCP09 uses low, medium, and high emissions scenarios consistent with those of the IPCC.

Observed trends from UKCP09 include that the temperature in central England has already risen by about 1°C since the 1970s and sea surface temperature around the UK coast has risen by 0.7°C in the last 30 years. Over the last 250 years, there has also been a slight trend for increased rainfall in winter and decreased rainfall in summer, as well as an increase in the intensity of winter rain.

The UKCP09 forecasts or projections are presented conveniently in a series of maps, two of which are reproduced in Fig. 6.2 showing the medium emissions scenario projections for the 2080s. Predictably, no part of the UK is cooler in the 2080s, but the south of England may become considerably warmer in the summer (around 3–4°C warmer). Summer rainfall in the 2080s is generally lower and sometimes quite a bit lower (perhaps almost a quarter lower in the south of England). Winter precipitation for the 2080s

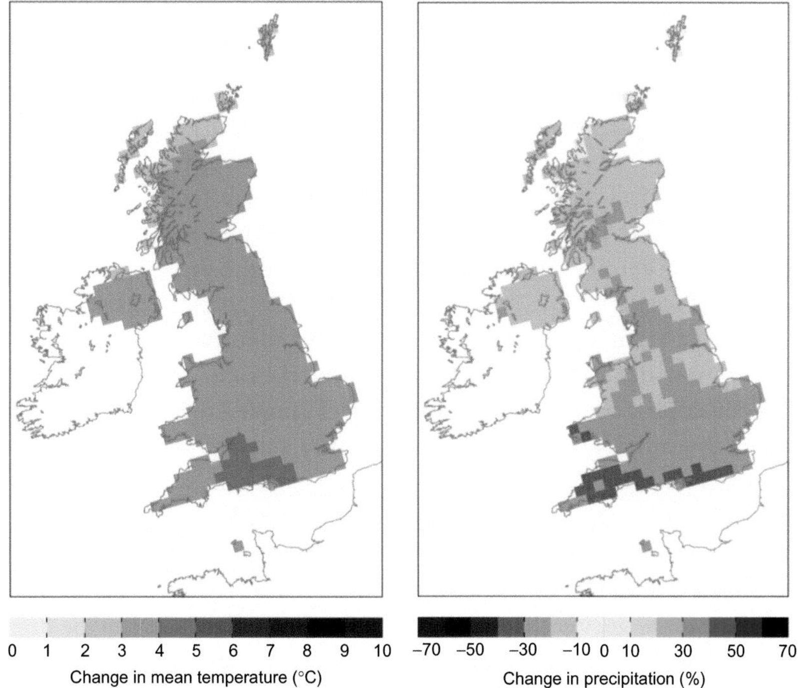

Fig. 6.2 Temperature and precipitation percent change from a 1961–1990 baseline for the 2080s for the medium emissions scenario. *(From Department for Environment, Food & Rural Affairs 2009. Adapting to climate change: UK Climate Projections, 52pp.)*

is generally higher over the whole of the UK. For the medium emissions scenario, sea-level rise is forecast at 18 cm in the 2040s and 36 cm in the 2080s.

Adaptations that might be needed include better cooling in buildings in the summer and better adaptation to winter flooding. For businesses there will be extra costs to adapt, though there may also be opportunities for new markets and new jobs. Agriculture will face changes in the growing season, droughts and floods, increased heat stress in livestock, increased possibility of fires, more storm damage, and increased risks from pests and diseases; but may also see increased yields in some crops, and the opportunity to grow new crops. In infrastructure, road and rail may need be to be rerouted and be proofed against higher groundwater levels, temperature, and rainfall. Sea defences will have to be strengthened and harbours modified.

Information about recent climate change in the developing world is more dispersed. Data seem to suggest that large parts of Africa, Asia, and Latin America have seen temperature rises over the past 30 years generally within the range of 0.5°C to 1.0°C, although there are regions with larger increases (for example in southeastern Brazil and northern Asia). Land surface air temperatures have risen by about double those over the ocean, with the result that less warming has occurred in small island developing countries (for example in the Pacific). There has, however, been a general increase in the frequency of warm extremes across the developing world.

Overall rainfall has decreased in the tropics since the 1970s. Regional variations include increased rainfall in eastern parts of South America and northern and central Asia, and reduced rainfall in the Sahel, southern Africa, and parts of southern Asia, leading to more intense and longer droughts. As elsewhere, there has been an increase in high-intensity rainfall events.

The most destructive of storms, tropical cyclones, disproportionately affect developing countries. Observed changes suggest that tropical cyclones show an upward trend in duration and intensity, related to higher tropical sea surface temperatures. The largest increase in the intensity of tropical cyclones is evident in the North Pacific, Indian, and southwest Pacific oceans.

There are fewer detailed projections of future climate change for the developing world than the developed world, but in general those that have been done show greater than mean level warming of land areas in the developing world because of lower availability of water for evaporative cooling. Increased rainfall is predicted in parts of the tropics, but overall reduced rainfall in the subtropics. Monsoons, which affect large parts of the developing world, will yield higher rainfall.

Africa stands out as a continent that will warm above the global annual mean throughout the continent and in every season. The drier subtropical parts of Africa will warm more than the wetter tropics. Annual rainfall will reduce in much of Mediterranean Africa, northern Sahara, and in southern Africa. The annual rainfall of eastern Africa will likely be higher. Changes in rainfall in already crucially stressed areas such as the Sahel and the southern Sahara are still uncertain.

In Asia, warming is predicted to be greatest in the continental interior and rainfall is predicted to increase in northern Asia, East Asia, South Asia, and most of Southeast Asia, but to decrease in central Asia. An increase in the frequency of intense precipitation events in South Asia and East Asia is associated with an increase in the intensity of tropical cyclones. Annual rainfall is projected to decrease in most of Central America but changes in annual and seasonal rainfall over northern South America, including the Amazon forest, remain uncertain. The climate in large parts of central and South America is affected by the El Niño Southern Oscillation (ENSO), but again the future behaviour of ENSO is uncertain.

CLIMATE CHANGE AND GROUNDWATER

The most important way that geological science can contribute to climate change adaptation is through understanding groundwater. A recent World Bank report states that the impacts of climate change will be channelled primarily through the water cycle, in that systems of food, energy, and urban and rural life will mainly feel the effects of climate change through water. In many parts of the world, groundwater is the chief source of water for domestic, agricultural, and industrial use, which is why the management of water (including groundwater) is so important, including through planning and regulation, pricing, and permits. But management of this type needs a deep understanding of the science of hydrology and hydrogeology.

In this section I will concentrate on hydrogeology, essentially the climatic factors affecting groundwater. Although groundwater receives little attention in policy discussion of climate change, it has a crucial role in providing a natural buffer against climate variability. Many of the most vulnerable areas, whether in developed or developing countries, abstract water from aquifers, particularly when surface runoff declines. Around 30% of global available fresh water resides in aquifers. About 70% of drinking water in the European Union, 80% of rural water supply in sub-Saharan Africa, and 60% of agricultural irrigation in India depend on groundwater. Many countries, therefore, have large groundwater-dependent economies. Groundwater also sustains ecosystems and landscapes in humid regions by supporting wetlands and other aquatic ecosystems.

The fact that groundwater is 'out of sight' means its development is often uncontrolled and not incorporated into overall river basin management, resulting in overexploitation and contamination. Like surface water, groundwater is also transnational in the sense that aquifers span international borders. Fig. 6.3 illustrates the extent to which large populations, particularly in the developing world, sit within the catchments of very large rivers. The politics of surface water that passes through borders can be fraught, but those of groundwater are largely undiscussed.

A few assertions can be made about the effects of climate change on groundwater. It is likely that recharge patterns will change. (Recharge is the process of water entering an aquifer mainly from the surface of the Earth). There is also likely to be increased demand, especially from irrigation, which today takes 70% of global groundwater withdrawals. Where total runoff is expected to decrease through declining precipitation, groundwater resources are also likely to decline. Changes in snowmelt will also affect groundwater recharge. Climate change may affect the quality of water

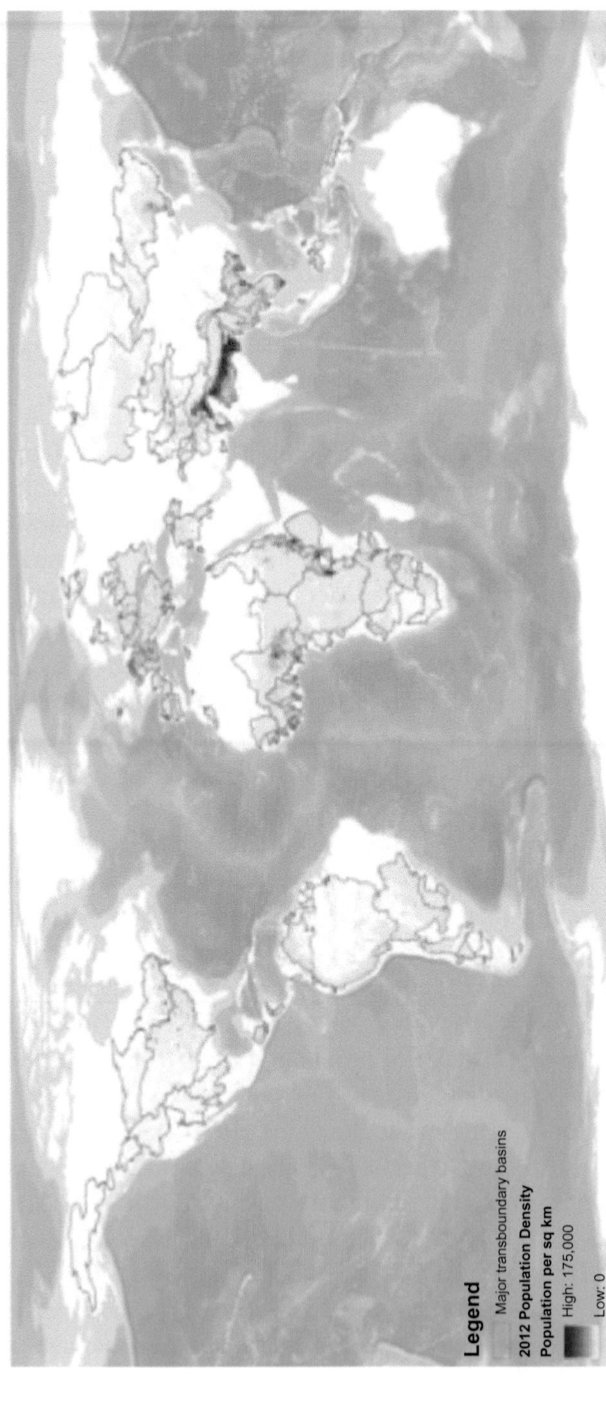

Fig. 6.3 Large transnational river basins and populations mean that surface water and groundwater are political issues. *(From World Bank Group, 2016. High and Dry: Climate Change, Water, and the Economy. World Bank, Washington, DC. 69pp.)*

in aquifers: with increasing temperature, groundwater salinity may increase as more water evaporates before it can reach deeper levels. Rising sea levels will also force seawater inland, changing recharge patterns.

However, if protected and managed along with surface water, groundwater can do much to help human populations to adapt to climate change. Its widespread availability and typically large volumes make it more naturally buffered against seasonal variations in rainfall and temperature. Unlike surface storage, aquifers lose negligible amounts of water through evaporation and transpiration.

Climate and Groundwater in the UK

Groundwater is a significant component of public water supply and water use in the UK, as well as sustaining rivers and wetlands. Across England and Wales the average annual recharge to the main aquifers is about 7 billion m^3. About a third of this is abstracted from aquifers at a rate of approximately 7 million m^3 per day.

Most of this groundwater is abstracted in southern, eastern, and central England from rock layers of Cretaceous chalk, Permo-Triassic sandstone, Jurassic limestone and Cretaceous 'Greensand'. Locally in the south of England groundwater may provide more than 70% of public water supply.

Given the present importance of UK groundwater, it is surprising that little is known about the effects of recent climate change on groundwater, nor the future effects. The reasons for this uncertainty include the long lag time between cause and effect, mainly because of the size of groundwater storage capacity, but also the complexity and heterogeneity of groundwater systems, including the rocks that host groundwater, land cover, land use, and water resource management.

A paper by Chris Jackson and colleagues reviewed changes in groundwater level (the level of the water table) in historical data. So-called groundwater droughts where water tables are low enough to cause abstraction problems are not difficult to identify, but statistical connections between groundwater levels and trends in global ocean and atmospheric circulation—in other words, a connection between observed climate change and groundwater level—are difficult to make.

Perhaps the most important aim is to establish a connection between observed groundwater drought and climate change so that some attempt can be made to predict how climate change might force future groundwater drought. Jackson and his colleagues identified trends in groundwater level data monitored in seven observation boreholes in the chalk aquifer over as much as a century since about 1900, and identified statistically significant trends of declining groundwater level at four of the sites (Fig. 6.4).

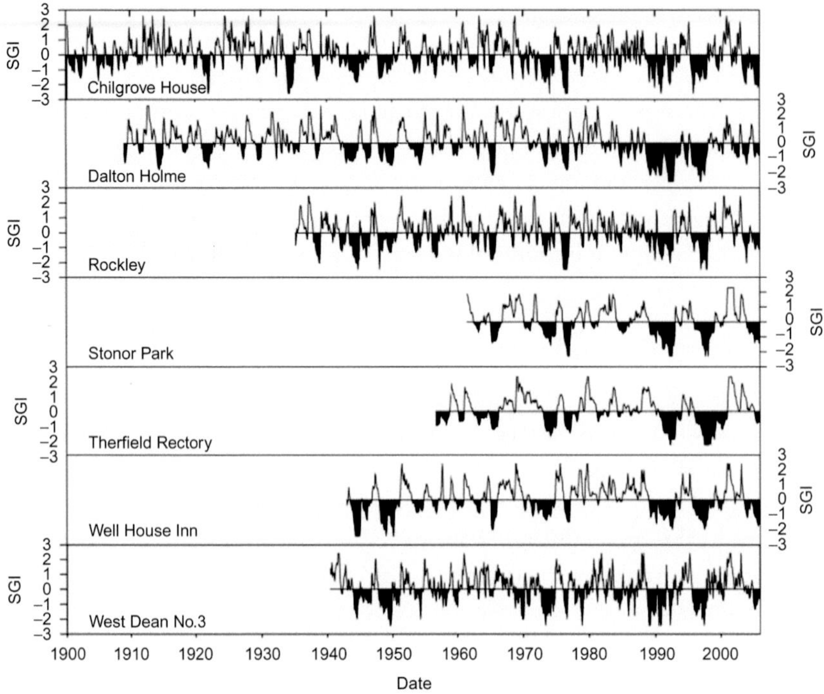

Fig. 6.4 Groundwater in seven boreholes in the Chalk aquifer of England. Statistical surveys showed declines in groundwater level in four of the sites. *(From Jackson, C. R. et al., 2015. Evidence for changes in historic and future groundwater levels in the UK. Progr. Phys. Geogr. 39, 49–67.)*

Further analysis into the future (out to the 2050s) suggests reductions (at most of the sites modelled) in annual and average summer groundwater levels and increases in average winter groundwater levels, under a high greenhouse gas emissions scenario.

Climate and Groundwater in Africa

Groundwater is probably even more important in Africa. There, groundwater is the major source of drinking water and its use for irrigation will probably increase as Africa's economy and population grow. Despite its importance, there is a dearth of detailed quantitative information on groundwater in Africa. Often groundwater storage is not factored in to assessments of freshwater availability.

An exception is a recent compilation of data by Alan MacDonald and colleagues that has allowed continent-wide estimates of the amount of groundwater and potential borehole yields, resulting in a figure for total groundwater storage in Africa of 0.66 million km^3 (Fig. 6.5). Not all of

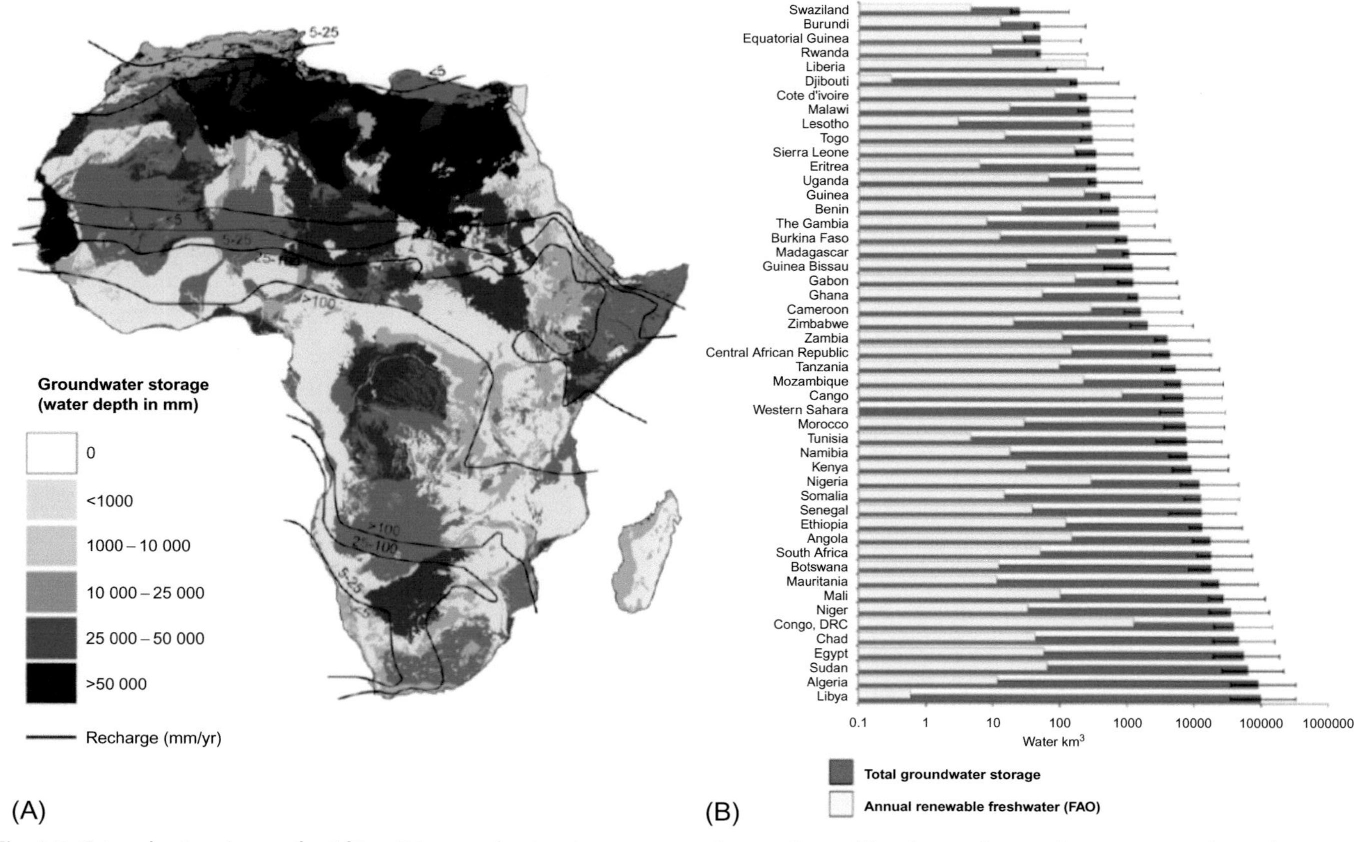

Fig. 6.5 Groundwater storage for Africa. (A) groundwater storage across the continent. (B) volume of groundwater storage for each country. *(From MacDonald, A.M. et al., 2012. Quantitative maps of groundwater resources in Africa. Environ. Res. Lett. 7, 024009.)*

this groundwater would be available from wells, but this figure of two-thirds of a cubic kilometre is 100 times larger than some estimates of annual renewable freshwater resources in Africa. This groundwater resource is not evenly distributed: most is present in North Africa (Libya, Algeria, Egypt, and Sudan). The same study indicates that, in many areas, well-sited and well-constructed boreholes could yield useful amounts of water for low-intensity rural activities, and the aquifers they penetrate will have enough water to sustain abstraction through seasonal variations (Fig. 6.6).

For industry and irrigation, the potential for higher-yielding boreholes, for example those that could deliver more than 5 L/s (litres per second), is much more limited. With climate change and population increase this is likely to pose a problem.

It is vital to understand the natural environment where there are rapidly growing populations and economies and where climate change is imminent. Obviously, this is a particular challenge for developing countries in Africa.

In East Africa, population is forecast to grow from about 300 million today to 800 million by 2060 and 1150 million by 2090. Many of the countries in East Africa are already water stressed based on per capita water availability, but by 2100 this will increase to nearly all. A review by a team led by Umesh Adhikari of the effect of climate change on runoff shows huge uncertainty in scientific studies of river catchments (Fig. 6.7, Table 6.1). An example is the upper Nile catchment, which is predicted to have a change by 2075 of between 25% less, or 32% more, runoff. Clearly a prediction involving less runoff has serious implications for the population that depends on water. How much of the deficit can be compensated for by, for example, groundwater? Are the aquifers large enough in the area and will water wells deliver the deficit? Will recharge in the area maintain the sustainability of the aquifer and the water wells or will it be affected by climate change in the area? Similarly, what would the effects be of a large increase in runoff? What measures might have to be put in place to deal with flooding? Much of the research needed to answer these questions locally has still to be done.

Another aspect of groundwater change in relation to climate change is saltwater intrusion. This is the movement of saline water into freshwater aquifers, which happens naturally in most coastal aquifers, because both kinds of groundwater are in close proximity. Coastal aquifers provide groundwater for more than a billion people. Saline groundwater is denser than fresh groundwater and so it tends to form wedge-shaped intrusions under freshwater. Intense abstraction from freshwater wells can draw saline water levels up and allow saline groundwater to penetrate further inland, below ground. Sea level rise due to climate change and increased marine

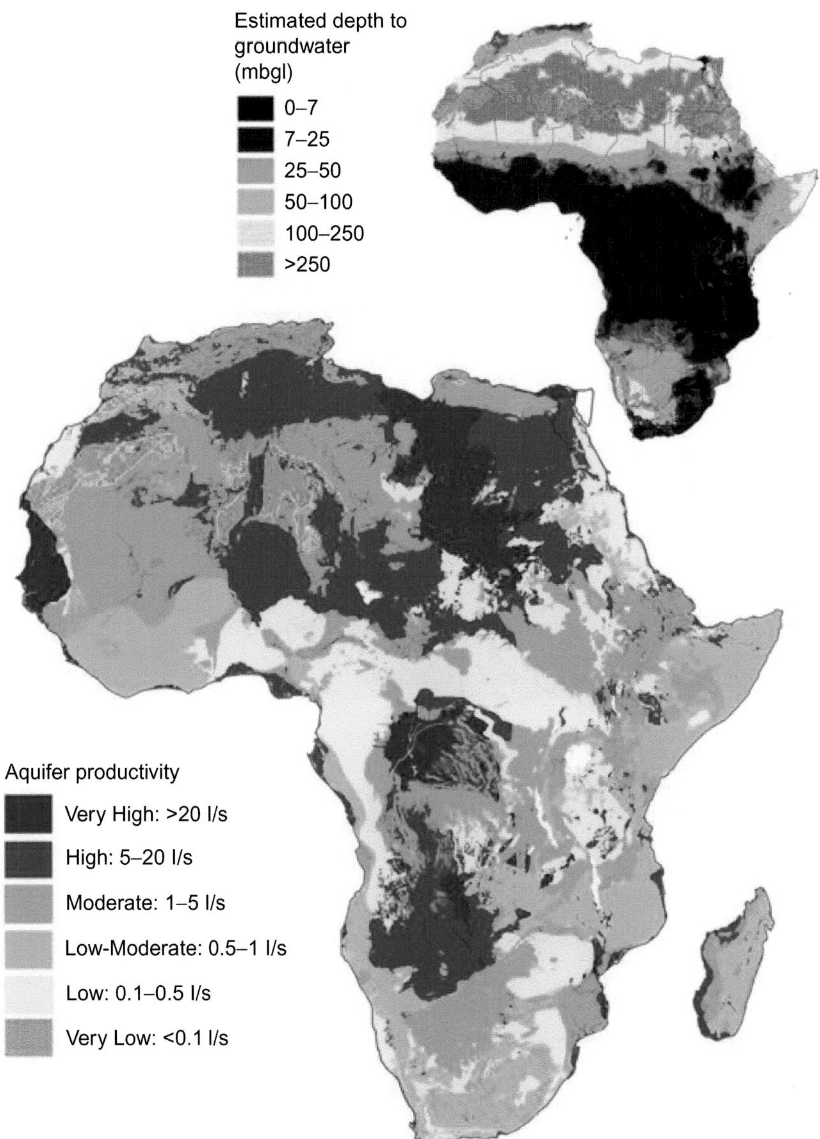

Fig. 6.6 Aquifer productivity for Africa. The inset shows approximate depth to groundwater. *(From MacDonald, A.M. et al., 2012. Quantitative maps of groundwater resources in Africa. Environ. Res. Lett. 7, 024009.)*

flooding that will result from climate change will likely increase saline groundwater influence in coastal areas.

Saline groundwater is not just a direct problem for domestic water and irrigation. Soils can also be affected. In Bangladesh, for example, increase in

Fig. 6.7 East African catchments and countries. *(From Adhikari, U., Nejadhashemi, A.P., Herman, M.R., 2016. A review of climate change impacts on water resources in east Africa. Trans. ASABE 58 (6), 1493–1507, Copyright 2016 American Society of Agricultural and Biological Engineers, used with permission.)*

soil salinity may lead to decline in yield of staple crops like rice and reduce the income of farmers significantly.

There are few detailed local answers to the many questions about runoff, inland groundwater, and coastal groundwater. Studies like those reviewed by Adhikari for East Africa are broad brush and lacking in local detail. The detailed study, when it comes, will have to be directed where

Table 6.1 Selected East African catchments and forecasted runoff change. The location numbers refer to Fig. 6.7

Location	Catchment	Country	Forecast target	Change in runoff
1	Lake Tana	Ethiopia	2075	−11.3%
2	Upper Nile	Ethiopia	2050	−25% to +32%
3	Nzoia catchment	Kenya	2050	+6% to +115%
4	Sezibwa catchment	Uganda	2100	+47%

From Adhikari, U., Nejadhashemi, A.P., Herman, M.R., 2016. A review of climate change impacts on water resources in east Africa. Trans. ASABE 58 (6), 1493–1507. Copyright 2016 American Society of Agricultural and Biological Engineers, used with permission.

population will likely concentrate in the future, and this needs an understanding of the way that large-scale development might occur, which in turn relates to the distribution of resources and already-established infrastructure. I will consider this later in the chapter.

Groundwater Adaptation

The role that groundwater could play in providing backup for reduced surface water availability or in conditions of greater seasonal variation needs research. This would involve mapping local and regional aquifers, knowing the effects of climate change on them, and having an idea of the human activities that they may have to support. This will involve science but also appropriate policy, regulation, and management. The range of adaptation measures might include deepening of existing boreholes, encouraging altered groundwater use and farm irrigation practices, and policy to protect groundwater. A summary of some of the actions that could be taken is shown in Table 6.2.

It is important to realise that many adaptations shown in the table, though effective locally, may have 'downstream' effects—in the sense of the effects of building dams on downstream river discharge, for example; or the effects on regional aquifers of increasing depth of water wells or their abstraction efficiency. River catchments and regional aquifers often cross international borders, which introduces a political element.

Less important in Africa but an important factor in Asian developing countries is the effect of increasing glacial melting in high tropical mountains, such as the Himalayas. This will mean increased river flows for a time. As glaciers retreat, glacial lakes sometimes form behind moraine or natural ice dams. These dams are comparatively weak and can breach suddenly, leading to a discharge of huge volumes of water and debris downstream, with effects on settlements, forest, farms, and infrastructure.

Table 6.2 Types of adaptation for water supply and demand

Supply side	Demand side
1. Increase storage capacity by building reservoirs and dams 2. Desalinate seawater 3. Expand rainwater storage 4. Remove invasive nonnative vegetation from riparian areas 5. Prospect for and extract groundwater 6. Develop new wells and deepen existing wells 7. Maintain well condition and performance 8. Develop aquifer storage and recovery systems 9. Develop conjunctive use of surface water and groundwater resources 10. Develop surface water storage reservoirs filled by wet-season pumping from surface water and groundwater 11. Develop artificial recharge schemes using treated wastewater discharges 12. Develop riverbank filtration schemes with vertical and inclined bank-side wells 13. Develop groundwater management plans that manipulate groundwater storage, e.g., resting coastal wells during times of low groundwater levels 14. Develop groundwater protection strategies to avoid loss of groundwater resources from surface contamination 15. Manage soils to avoid land degradation to maintain and enhance groundwater recharge	1. Improve water-use efficiency by recycling water 2. Reduce water demand for irrigation by changing the cropping calendar, crop mix, irrigation method, and area planted 3. Expand use of water markets to reallocate water to highly valued uses 4. Promote traditional practices for sustainable water use 5. Expand use of economic incentives including metering and pricing to encourage water conservation 6. Introduce drip-feed irrigation technology 7. Licence groundwater abstractions 8. Meter and price groundwater abstractions

After Green, T.R., 2011. Beneath the surface of global change: Impacts of climate change on groundwater. J. Hydrol. 405, 532–560.

Groundwater and Geomicrobiology

Perhaps a final mention is needed of the role of microorganisms in the subsurface and how they might affect global biogeochemical cycles and climate change. It is well established that aquifers contain microorganisms and that these are generally connected with disease pathogens in drinking water or subsurface engineering challenges such as pipeline corrosion or blockages in oil and gas reservoirs. However, the natural ecology of groundwater is also important in sustaining water quality, and can help to restore contaminated aquifers.

The role of subsurface microorganisms in the carbon cycle is largely unknown. Microbial processes and diversity in aquifers are not well understood. Climate change could, for example, lead to changes in the metabolic rates of aquifer microorganisms, altered ecosystem productivity and biodiversity, and changes in species ecology. Overabstraction as a response to climate change could increase the number of pathogens in aquifers, for example *Cryptosporidium*, which can cause respiratory and gastrointestinal illness. The contributions to global biogeochemical cycles that aquifer microorganisms undoubtedly make (for example gases produced in metabolism) are probably closely climate sensitive, and the effects of changing them are mainly unknown.

THE FOOD-WATER-ENERGY NEXUS

It is clear that food, water, and energy are interconnected and are in tension. This is embodied in the concept of the food-energy-water nexus (Fig. 6.8). Water is used for energy activities such as oil and gas drilling and refining, as well as for growing biofuel crops and for generating electricity in most sorts of power stations. The multiple interactions involve space for biofuels that

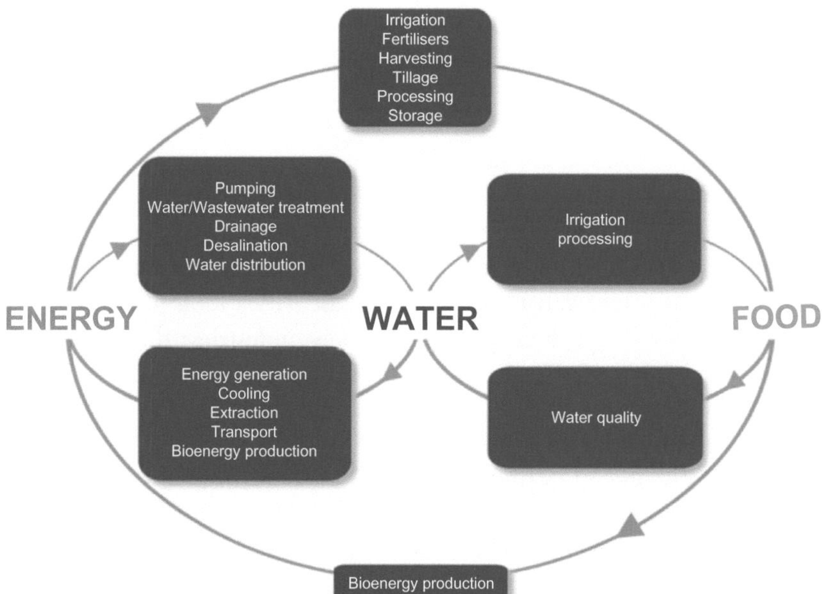

Fig. 6.8 One representation of the food-water-energy nexus. *(From the International Renewable Energy Agency, 2015: Renewable Energy in the Water, Energy & Food Nexus.)*

could be used for food, and water for energy generation that could be used to grow food crops.

Energy activities also produce waste water either from the processes themselves or because, in the case of oil and gas for example, drilling liberates deep, highly saline water. Such waste water may be too polluted, saline, or warm to be easily disposed of without careful treatment. The treatment itself of course will require energy. Energy is also required to pump groundwater, or desalinate seawater.

Last but not least, food production consumes more fresh water than any other human activity. Agriculture is responsible for an average of 70% of fresh water consumption, and in some countries the figure is 80%–90%. Food production also affects water because it involves land-use change, including changes in runoff and groundwater discharge. Modern agriculture and fertilisation cause groundwater pollution (for example nitrate pollution). Agriculture also consumes a lot of energy. Agriculture and food and the associated supply chains use up 30% of total global energy, and this is manifested in the link between food prices and the prices of primary fuels such as oil.

Mathematical modelling can be used to look at parts of the problem, for example the links between water and food. Modelling of water use and crop yield by a group based at the Australian National University for 19 countries out to 2050 showed that, though there may be enough crop-producing land, there will not be enough water for irrigated agriculture to provide food in several large food-producing countries. These tensions could be severe, given that world food demand is likely to increase by around 60% by 2050 and that irrigation presently accounts for 70% of global water need, and around half of food production.

Another way to look at the complex links and limits within the food-water-energy nexus, particularly in developing countries where the pressure will be the most severe, is to identify where future infrastructure development will take place. What will India or East Africa look like in 2060 or 2090? Where will the new people, the new industries, the new towns and cities be? It may also be worth looking at how African resources developed in the past.

A good example of past development is the Zambian copper belt. Copper began to be mined in the copper belt in the 1920s in areas of virtually empty countryside. Within a few decades the copper belt contained several of Zambia's largest towns, Kitwe, Ndola, Mufulira, Luanshya, Chingola and Chililabombwe—and at about 2 million people, 15% of its population. The area now also has a legacy of poor environmental management and land degradation related to mining. The broader environmental problems of the copper belt include deforestation, unplanned urban development, and eutrophication of waterways by sewage effluent.

Degradation directly related to mining itself includes air pollution, particularly in the form of SO_2 from copper smelters which converts to sulphuric acid, causing soil and vegetation pollution. Nitrous oxides and organic acids also enter streams and affect aquatic fauna. Direct contamination from chemical and oil spills has also occurred. Runoff and leakage from waste dumps and tailings dams pollute streams flowing out of the mine areas. Occasionally, sudden failure of tailing dams causes extensive physical and ecological damage. Lead contamination in soils is also a problem.

How do we prevent the same pressures, clearly related to the food-water-energy nexus, from happening again? What can we learn from developments like the copper belt? A good guide might be the pattern of potential resources like land, water, and minerals, as well as existing infrastructure and transport routes. These items are linked into arcs or strips of land known as development corridors. It is in these development corridors where the food-energy-water nexus will be most critical and need the most management.

Whatever the resources that the development corridor might link, their growth will ultimately be driven by economics. A development corridor might start as a basic transport route and the addition of other types of transport produces a transport corridor. Efficient corridor operations encourage further economic activity that leads to further investment and, ultimately, the corridor evolves into an 'economic corridor' (Fig. 6.9).

It is worth looking at a couple of development corridors in Africa to illustrate the point. The so-called 'Northern Corridor' already links the landlocked countries of Uganda, Rwanda, and Burundi with Kenya's port of Mombasa (Fig. 6.10). The Northern Corridor could, with more development, serve the eastern part of the Democratic Republic of Congo, Southern Sudan, and northern Tanzania.

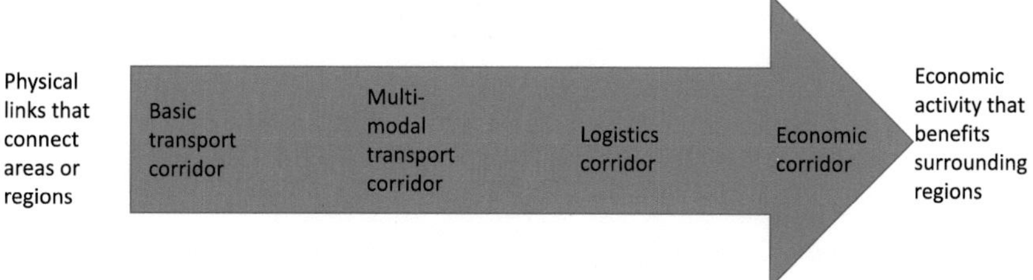

Fig. 6.9 How a transport corridor evolves into an economic corridor. *(Adapted from Hope, A., Cox, J., 2015. Development Corridors. Coffey International Development, 74pp., Fig. 4.)*

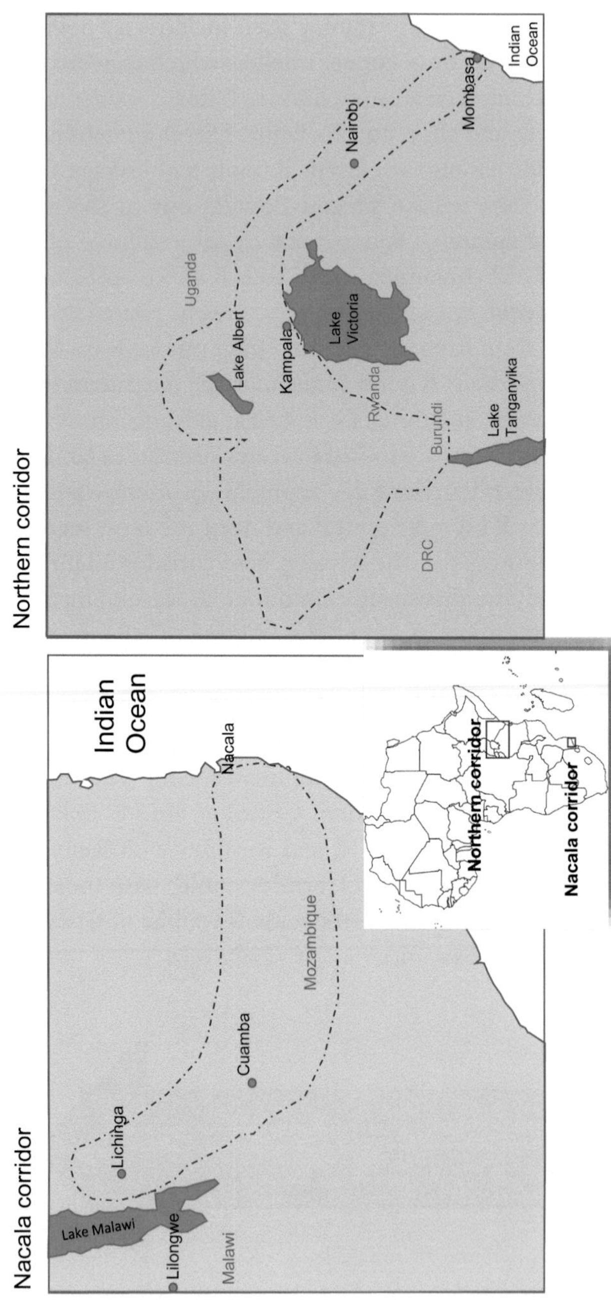

Fig. 6.10 The Nacala and Northern corridors. *(Adapted from Hope, A., Cox, J., 2015. Development Corridors. Coffey International Development, 74pp., Maps 8 and 9.)*

The Nacala Corridor to the south is less developed. There is considerable governmental and commercial interest because its future purpose would be to unlock the development potential of the hinterland of the Nacala Port and rather underdeveloped parts of Mozambique, Malawi, and Zambia. New resources will be accessible economically and the development of business and commerce will contribute to the reduction of poverty.

The likelihood is that the pattern of development corridors either planned or already in existence is a good guide to the concentration of population and industry, which is the key to understanding the food-water-energy nexus, and the geological aspects of the nexus, whether it be well-managed groundwater inland or at the coast.

Recent thinking on the economic system of the food-energy-water nexus can help to see the natural parts as economically linked through the concept of 'natural capital'. Natural capital is the value of the world's stock of natural assets, including geological resources, soil, air, water, and all living things from which humans derive a wide range of services, often called ecosystem services. A simple explanation of the principle of natural capital was provided by Dieter Helm in his recent book *Natural Capital*. In the book, Helm describes how much we invest in crucial infrastructure such as electricity and gas transmission systems and asks why we do not invest to the same extent in, for example, natural bee pollination and natural salmon fisheries. After all, are these resources not just as vital to our future survival as electricity and gas?

In the UK, the Natural Capital Committee was set up to advise the Government on these issues, and it recently recommended an approach related to river catchments. The concept is that countries can be divided into their large river catchments (bordered by their major watersheds), and that these river catchments are essentially resource/ecological zones or 'natural capital' building blocks. The catchments may cross internal UK political boundaries but are more sensible in that they are more natural units. Each catchment could have a natural capital 'account' managed locally, including a natural environmental baseline, a beneficiary analysis, and a delivery plan. Attempts are being made at how this approach might work, for example in the Poole Harbour catchment in Dorset in the south of England, and, internationally, in the cross-border catchments of the countries of the Southern African Development Community (SADC).

URBAN DRAINAGE AND CLIMATE CHANGE

Groundwater is an important pool that protects against climate change, which reduces surface water availability. But the reverse—an excess of surface fresh water—can cause problems, particularly in urban areas, especially as more absorbent rural land is covered by the impermeable concrete of expanding cities and as saline water pushes underneath freshwater coastal aquifers, raising groundwater levels. Simple adaptations will involve waterproofing electrical lines and communications cables or locating critical infrastructure away from low-lying ground, and proofing buildings against higher groundwater levels.

But what can be done to reduce storm water getting in the way of infrastructure?

For sudden increases in runoff, cities can have concrete channels and conduits to take storm water out to sea as soon as possible. They can also try to encourage water to sink harmlessly away. 'Tree trenches', for example, are ditches excavated along city pavements filled with gravel and soil and planted with trees that help to take water away from the surface—and also improve the appearance of urban streets (Fig. 6.11). Permeable asphalt can

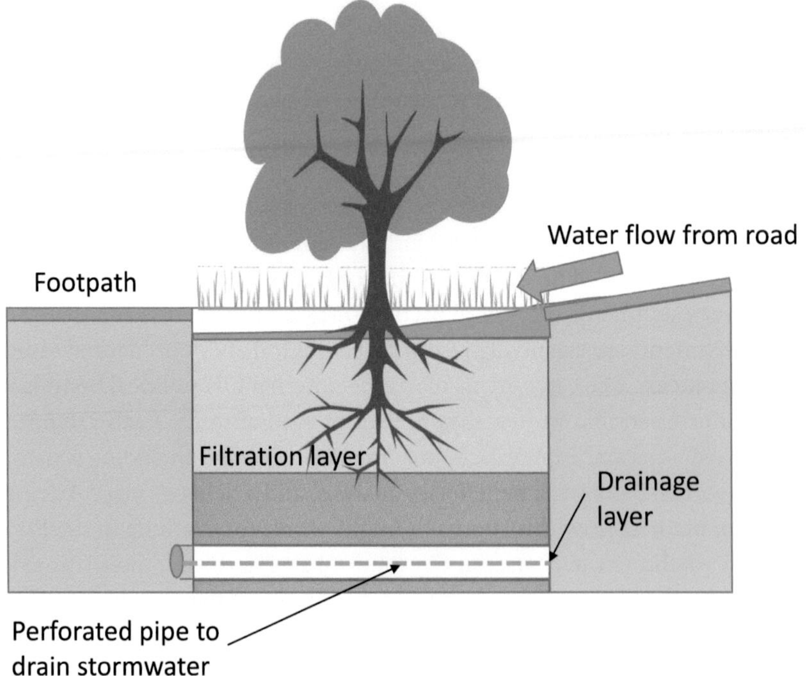

Fig. 6.11 'Tree trenches' are ditches excavated along city pavements filled with gravel and soil and planted with trees that help to take water away from the surface. *(Redrawn from http://urbanwater.melbourne.vic.gov.au/projects/raingardens/little-collins-street-tree-pits/.)*

also be used rather than the impermeable variety, which allows water to soak through the road and pavement, reducing the flow in city drains.

Some cities are trying to increase the reservoir capacity of water storage in engineered subsurface tanks. These storm-water reservoirs allow water to collect and be used as a resource after purification.

SUMMARY

This chapter has shown in summary the recent changes attributed to climate change, and forecasts for future climate, derived from mathematical modelling. These reviews and forecasts are easier to make for large areas but lose their precision with increasing focus on the local; and 'the local' is the place where climate change will be felt most. Planners and policymakers are faced with difficult decisions about cost-effective adaptations. In geological terms, perhaps the most important 'carrier' of climate change is groundwater. In many ways groundwater has the potential to be very useful in providing backup for surface water when its availability is in question, and so is a buffer for climate change. The problem is that it is still not exactly clear how climate change will affect the amounts of groundwater, nor, for example, the microorganisms in groundwater. In Africa, groundwater is clearly vital' but over large areas of the continent, water wells will not be able to provide the sorts of flow that are needed to sustain agricultural and industrial development. There is also uncertainty about the precise locations of the most intense future development in Africa. The positions of present and future 'development corridors' may allow greater understanding of the food-water-energy nexus in Africa.

Most of this chapter has been about the ground as a storage place for groundwater and the role of water in protecting against climate change that reduces surface water availability. In fact, the ground also has a strong role in guarding against climate change that produces an excess of water, particularly in cities. This includes using purpose-built underground reservoirs to contain storm water and encouraging infiltration into the permeable rock and soil of the city through permeable pavements and tree trenches.

BIBLIOGRAPHY

Adhikari, U., Nejadhashemi, A.P., Herman, M.R., et al., 2016. A review of climate change impacts on water resources in east Africa. Trans. ASABE 58 (6), 1493–1507.

Barbosa, A.E., et al., 2012. Key issues for sustainable urban stormwater management. Water Res. 46, 6787–6798.

Bloomfield, J.P., Marchant, B.P., 2013. Analysis of groundwater drought building on the standardised precipitation index approach. Hydrol. Earth Syst. Sci. 17, 4769–4787.

Christensen, J.H., et al., 2007. Regional climate projections. In: Solomon, S., Qin, D., Manning, M., Chen, Z., Marquis, M., Averyt, K.B., Tignor, M., Miller, H.L. (Eds.), Climate change 2007: the physical science basis. Contribution of Working Group I to the fourth assessment report of the intergovernmental panel on climate change. Cambridge University Press, Cambridge, pp. 847–940.

Conway, D., et al., 2015. Climate and southern Africa's water-energy-food nexus. Nature Climate Change 5, 837–846.

Darby, S.E., et al., 2016. Fluvial sediment supply to a mega-delta reduced by shifting tropical-cyclone activity. Nature. 2016 Oct 19, https://doi.org/10.1038/nature19809 (Epub ahead of print).

Dasgupta, S., et al., 2014. River Salinity and Climate Change: Evidence from Coastal Bangladesh. Policy Research Working Paper 6817, The World Bank Development Research Group Environment and Energy Team. 44pp.

Department for Environment, Food & Rural Affairs, 2009. Adapting to climate change: UK Climate Projections. 52pp.

Gill, J.C., 2017. Geology and the sustainable development goals. Episodes 40, 70–76.

Grafton, R.Q., et al., 2017. Possible pathways and tensions in the food and water nexus. Earth's Future 5, 1–14.

Green, T.R., 2011. Beneath the surface of global change: Impacts of climate change on groundwater. J. Hydrol. 405, 532–560.

Gregory, S.P., et al., 2014. Microbial communities in UK aquifers: current understanding and future research needs. Quart. J. Eng. Geol. Hydrogeol. 47, 145–157.

Helm, D., 2015. Natural Capital: Valuing the Planet. Yale University Press. 320pp.

Hope, A., Cox, J., 2015. Development Corridors. Coffey International Development. 74pp.

IPCC, 2014. Summary for policymakers. In: Field, C.B., Barros, V.R., Dokken, D.J., Mach, K.J., Mastrandrea, M.D., Bilir, T.E., Chatterjee, M., Ebi, K.L., Estrada, Y.O., Genova, R.C., Girma, B., Kissel, E.S., Levy, A.N., MacCracken, S., Mastrandrea, P.R., White, L.L. (Eds.), Climate Change 2014: Impacts, Adaptation, and Vulnerability. Contribution of Working Group II to the Fifth Assessment Report of the Intergovernmental Panel on Climate Change. Cambridge University Press, Cambridge, United Kingdom and New York, NY.

Jackson, C.R., et al., 2015. Evidence for changes in historic and future groundwater levels in the UK. Progr. Phys. Geogr. 39, 49–67.

Khan, M.R., et al., 2016. Megacity pumping and preferential flow threaten groundwater quality. Nature Communications 7, Article number: 12833, https://doi.org/10.1038/ncomms12833.

Mabhaudhi, T., et al., 2016. Southern Africa's water–energy nexus: towards regional integration and development. Water 8, 235.

MacDonald, A.M., et al., 2012. Quantitative maps of groundwater resources in Africa. Environ. Res. Lett. 7, 024009.

Mertz, O., et al., 2009. Adaptation to climate change in developing countries. Environ. Manage. 43, 743–752.

Papacharalampou, C., et al., 2017. Catchment metabolism: Integrating natural capital in the asset management portfolio of the water sector. J. Cleaner Prod. 142, 1994–2005.

Vd Dobblesteen, A., et al., 2012. Cities as organisms. In: Roggema, R. (Ed.), Swarming Landscapes: Advances in Global Change Research. 48. Springer, Netherlands, pp. 195–206.

Watts, G., et al., 2015. Climate change and water in the UK—past changes and future prospects. Progr. Phys. Geogr. 39, 6–28.

World Bank Group, 2016. High and Dry: Climate Change, Water, and the Economy. World Bank, Washington, DC. 69pp.

CHAPTER 7

Feedbacks and Tipping Points

Contents

The climate is a system like any other complex system in that it has regulatory forces that tend to keep things as they are, and elements that tend to cause change. These are negative and positive feedbacks. Study of feedbacks is very important to understand imminent climate change, and also to understand how strong positive or 'runaway' feedbacks can cause revolutionary (rather than 'evolutionary') change, where the Earth 'flips' from one state to another through a 'tipping point'. These forces are increasingly well understood, particularly in modern environments, for example the sea-ice loss effect which, revealing the darker ocean underneath, allows absorption of more heat, causing more ice to melt in a spiral of increasing temperature. Studies from deep time, for example from the Palaeocene-Eocene Thermal Maximum, might help us further understand these tipping points.

But it is also interesting that energy systems—complex systems of economics, technology, and geopolitics—also display similar forces, including serendipitous events leading to the increased use of fossil fuels, positive feedback that allows fuel use to rapidly grow, and tipping points that lead to completely new systems in energy. Perhaps the most obvious example of this kind of tipping point is the industrial revolution.

Knowing that these forces for change exist helps us also understand ways that dangerous feedbacks and tipping points can be avoided, paradoxically introducing a completely new 'teleological feedback'—that of conscious human intervention to prevent harmful change.

Despite the evidence of Chapter 2 of this book, which tends to concentrate on change in deep time, the overall history of the Earth's climate suggests very long-term stability on which are superimposed short periods of instability, such as the Palaeocene-Eocene Thermal Maximum (PETM) and the events at the Permian-Triassic boundary. As we have seen, these periods of instability result from geologically sudden changes in palaeogeography, greenhouse gas concentrations, incoming solar radiation, and ocean heat transport.

Energy and Climate Change
https://doi.org/10.1016/B978-0-12-812021-7.00007-5

The overall stability which keeps the Earth within the bounds of habitability was noted by James Lovelock and Lynn Margulis and became the basis of the Gaia hypothesis—broadly, that life (without conscious thought or purpose) has regulated the composition of the atmosphere, and therefore the climate, over many hundreds of millions of years. The carbon cycle is just one of the many biogeochemical cycles that are at the heart of this regulation that results from the complex interaction between the biosphere, atmosphere, and geosphere.

Such concepts of regulation and sudden change are familiar to engineers and mathematicians who work in systems theory. Essentially the climate system can be looked at as having regulatory aspects that tend to keep things as they are, and elements that tend to cause change. These are negative and positive feedbacks.

For example, in relation to atmospheric temperature, when a change in temperature causes an event to occur which itself changes global temperature, this is referred to as a feedback. If this feedback acts in the same direction as the original temperature change, it is a 'destabilising positive feedback', because warming is causing more warming. However, if it acts in the opposite direction, it is a 'stabilising negative feedback', for example where warming eventually causes a cooling effect. If there is a sufficiently strong net positive feedback, a 'revolutionary' change may occur where a system with inherent stability becomes another with a different kind of stability. At this point a climate 'tipping point' has been passed. Related to the tipping point is the idea of the 'runaway greenhouse effect' where climate change is at such a rate that it produces 'out-of-control' amplifying, positive feedbacks.

Going back to the PETM described in Chapter 2, there appears to be evidence for sudden big changes in the atmosphere. There is a large shift in the carbon isotope composition ($\delta^{13}C$) of carbon in sediments in around 130 locations in a wide range of ancient Palaeocene-Eocene environments, and a sudden absence of carbonate from Palaeocene-Eocene sediments of the deep sea, assumed to be related to ocean acidity.

Although the exact mechanism is unknown, the initial source of excess CO_2 may have been volcanic processes. The most obvious positive feedback that could have amplified this initial event is the destabilisation of marine methane hydrates. As we have already seen, under certain temperature and pressure conditions, methane—which is being produced continually by decomposing microbes in sea-bottom sediments—is stable in ice-like molecular cages trapping the methane in solid form. If seawater warms, for example in the early stages of global warming, seabottom methane hydrates will become unstable, releasing this potent greenhouse gas into the atmosphere to allow yet more comprehensive warming of the seabed and more release of methane. Some independent evidence that methane, though not necessarily methane hydrate, did play a part in the PETM is the size of the change in $\delta^{13}C$ of the carbon in PETM sediments, which is consistent with the very low $\delta^{13}C$ of biogenic methane in hydrates. In the terminology of climate systems, the positive feedback is clear. Whether a tipping point was reached at the PETM depends on your frame of reference. Clearly, big changes resulted, including faunal and floral extinction and widespread environmental change. It is interesting that temperature did eventually decrease, because the PETM probably lasted less than 100,000 years and the atmosphere did return to a more typical state. Paradoxically the most recent work suggests that this may have been due to an overall increase in biological productivity, in the form of a carbon pump transporting carbon from the atmosphere to the deep ocean. So, although changes were very extensive, a long-term carbon cycle process acted as a negative feedback in a regulatory way.

In fact, we know more about modern feedbacks in climate change. Clouds, for example, act as reflectors by bouncing back into space about one-third of the total amount of sunlight that hits the Earth. In simple terms, global warming increases the capacity of the atmosphere to hold water, leading to more clouds, reducing the sunlight that reaches the surface of the Earth. This is negative regulatory feedback. Increased rainfall and CO_2 could be similar in that they stimulate more plant growth and therefore more take-up of CO_2.

An example of a strong instability-causing, positive feedback is ice loss, particularly over the sea. Ice is very reflective in comparison to the sea's surface, which is dark and absorbs heat. Global warming melts sea ice, and the darker ocean absorbs more heat, causing more ice to melt in a spiral of increasing temperature (Fig. 7.1).

Changes in ocean composition or circulation could bring about abrupt shifts akin to tipping points. For example, as Arctic sea ice and Greenland ice sheets melt, large amounts of freshwater would affect ocean circulation in the Atlantic by reducing the northward flowing thermohaline ocean circulation and Gulf Stream, slowing the ocean 'conveyor belt' which brings warm water northward in the Atlantic Gulf Stream. Changes in the Southern Ocean would be particularly problematic because it is the world's biggest global CO_2 'sink', absorbing around 10% of anthropogenic CO_2 emissions

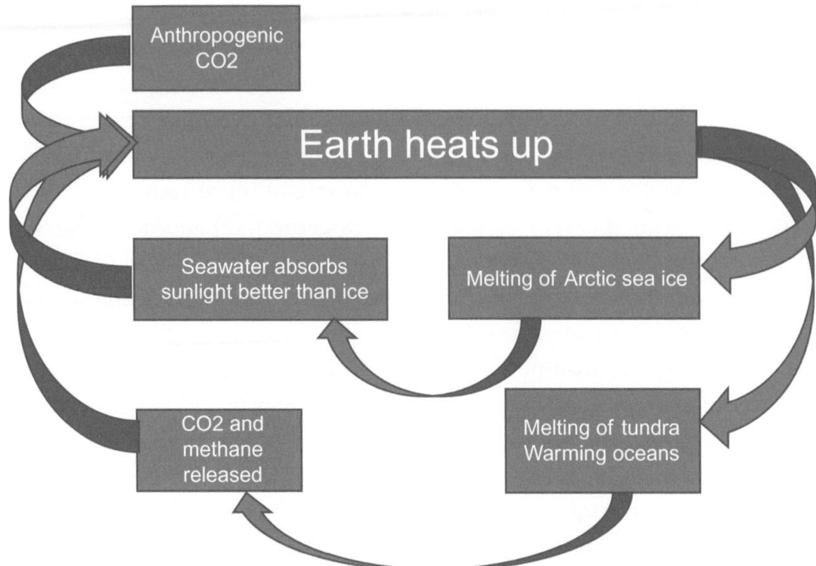

Fig. 7.1 The climate is a system like any other complex system in that it has regulatory forces that tend to keep things as they are, and elements that tend to cause change. *(Adapted from http://globalclimatechangenow.blogspot.co.uk/p/feedback-loops.html.)*

and storing them in cold bottom waters where CO_2 is retained far longer than it would be at the warmer sea surface.

The Amazon basin could also be the site of a tipping point. Rainfall there is largely recycled from water within the rainforest, but it is also influenced by wider patterns such as the El Niño Southern Oscillation (ENSO). If, as predicted, the Amazon basin becomes drier, diminished vegetation could reduce recycling of water to the rest of the rainforest in a positive feedback. The northern Hemisphere boreal or taiga forest of Canada and Russia is adapted to certain conditions of permafrost and fire. Drier summers could make the forest more susceptible to fire. The reduction of forest biomass as a whole across the Earth reduces the efficacy of the terrestrial carbon pump, allowing higher temperature and lower rainfall in a positive feedback.

The concept of the Anthropocene Epoch that I mentioned in an earlier chapter has concentrated minds on the effects that humankind has had on the climate system. Discussions are taking place about the best way to define the base of the Anthropocene Epoch, alongside all the other geological periods, in layers of rocks (Fig. 7.2).

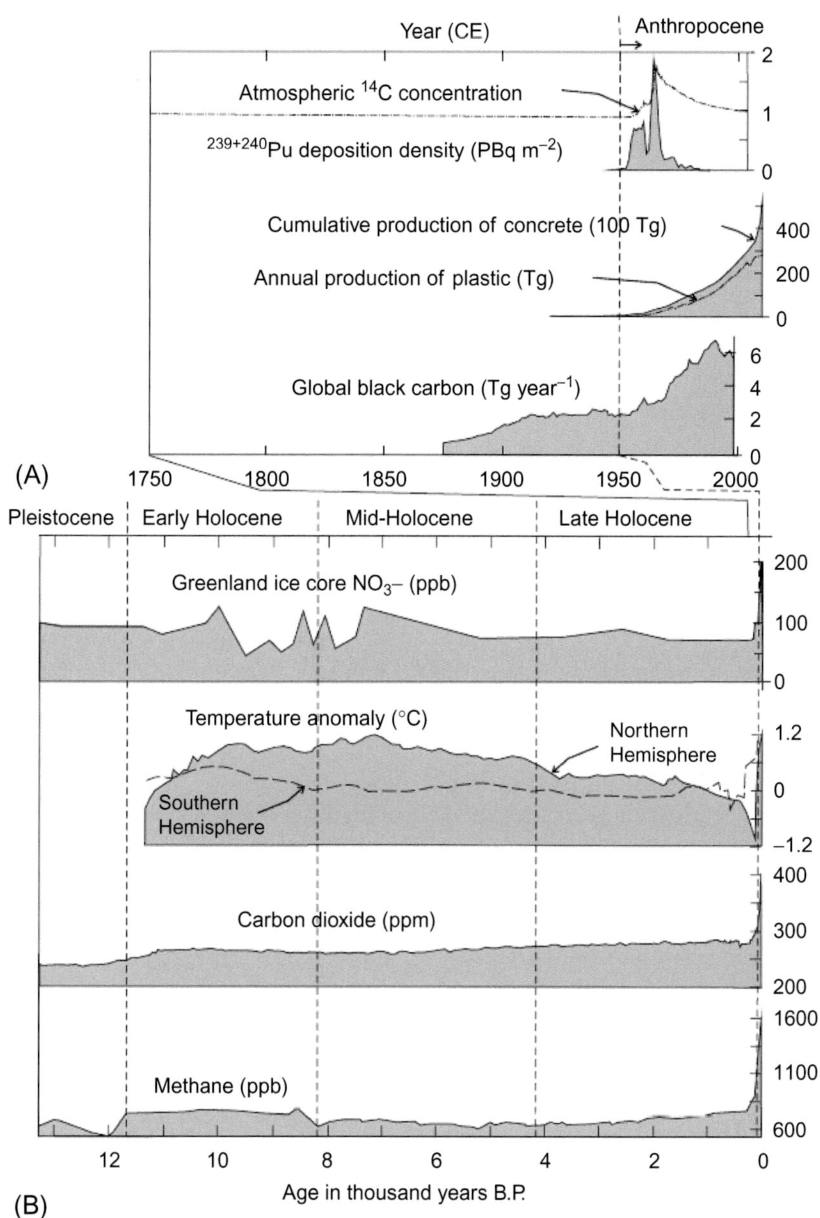

Fig. 7.2 Summary of key markers of anthropogenic change that could be used to mark the Anthropocene. (A) Markers, such as concrete, plastics, global black carbon, and plutonium (Pu) fallout, shown with radiocarbon (^{14}C) concentration. (B) Longer-ranging signals such as nitrates (NO_3^-), CO_2, CH_4, and global temperatures. *(From Waters, C. N. et al., 2016. The Anthropocene is functionally and stratigraphically distinct from the Holocene. Science 351, 137–147.)*

Notable geochemical changes that occur in sediments or ice that might be used to define the base of the Anthropocene include increases in polyaromatic hydrocarbons and pesticides, lead, ^{14}C, and plutonium. Changes in the ratios of lead isotopes connected with the uptake of leaded petrol appear around 1945. Perhaps the sharpest of the changes are of ^{14}C and ^{239}Pu, related to nuclear weapons testing, which had a global footprint between 1952 and 1980, known as the 'bomb spike'.

TIPPING POINTS AND FEEDBACKS IN HUMAN ECONOMIC AND CULTURAL CHANGE

The changes in the natural world shown in the upper part of Fig. 7.2 are a result of human economic and cultural change. Interestingly, these trends contain analogous events to the positive feedbacks and tipping points revealed by science. These particularly relate to energy and fossil fuels—including serendipitous events leading to escalations in the use of fossil fuels, positive feedbacks that allows fuel use to rapidly grow, and tipping points that lead to completely new systems in energy. Perhaps the most obvious examples are the energy transitions that I described in an earlier chapter: from wood to coal, from coal to oil, and from oil to gas or renewables, or some combination thereof.

In an earlier chapter I outlined the advent of the fossil economy in the industrial revolution and the way that it represented not just a change in the fuel that humankind uses, but also in the way that humankind became less restrained by the natural limits of the previous 'prime mover', that is, water.

What sorts of positive feedbacks allowed this change to happen so quickly and thoroughly? Perhaps the simplest example is the steam engine and its relationship with coal. During the industrial revolution, steam engines that burned coal replaced water and wind power, and became the dominant source of power from the late 19th century to the early 20th century, when steam turbines and internal combustion engines replaced steam engines. Despite this, steam is still with us in the sense that the steam turbine is still the most common driver of electrical generators in power stations. The key to the feedback though is that steam power could power pumps that allowed water to be removed from coal mines, to allow more coal to be mined. The first of this type of engine to be used in a coal mine was built by Thomas Newcomen in 1712 for a mine at Dudley Castle in Staffordshire, England. By 1725 this 'Newcomen engine' was in common use in coal mines. The better-known Watt steam engine, developed between 1763 and 1775, was an improvement on the Newcomen design.

Many other trends acted in the same way to stimulate more change, for example the development of steel, which depended on coal, the transport of which required railways, which in turn needed steel for rails. The ability to smelt lead using coal allowed the manufacture of lead-lined vessels, which allowed, for example, the manufacture of large amounts of sulphuric acid used in countless industrial activities. Sulphuric acid quickly substituted for natural methods of bleaching cotton in cotton mills, replacing natural methods (for example urine).

The same sorts of feedbacks could be said to have accelerated the rate at which oil and gas were taken up as fuels. The concentration of cheap gas in Pennsylvania helped Pittsburgh to develop as an industrial centre in the 1890s, particularly in steel. The development of industrial automobile production in the early years of the 20th century was a good use for steel and oil. Steel for rotary drilling equipment helped to hugely increase the production of oil between 1900 and 1920.

Analysis of energy transitions also indicates some of the difficulties of moving from one system of energy to another: the intrinsic inertia in systems which makes transitions more difficult, for example the transition from hydrocarbons to renewables. Transition begins with experimentation to small-scale technology and diversity of designs. This is followed by scale-up with associated economies of scale to industry level, and standardization. At this point, change becomes difficult because technology is frozen inside the system.

Fig. 7.3 illustrates this diagrammatically but is perhaps best envisioned as a physical model whereby a ball is installed in a trough. Where perturbations to the system are small, the ball can swing from side to side without the

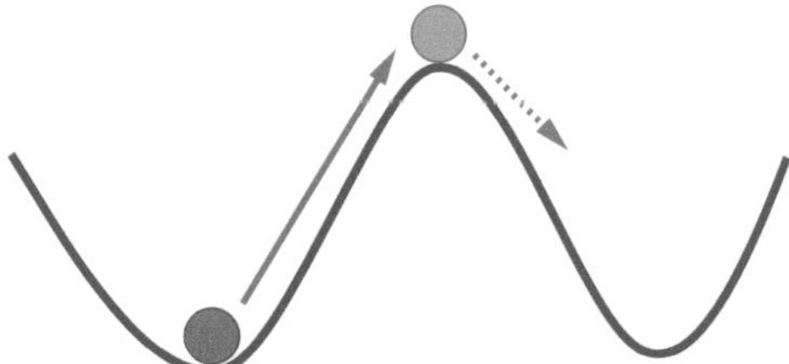

Fig. 7.3 A diagrammatic concept of the tipping point. If there is sufficiently strong net positive feedback, a revolutionary change may occur where a system with inherent stability becomes another with a different kind of stability. At this point a climate 'tipping point' has been passed.

'momentum' to reach the crest. This is a resilient system that absorbs shocks. Where positive feedback is strong enough, the ball can be propelled into the next trough, passing the tipping point.

Looking back to the IEA's view of the feasibility of the aims of the Paris Agreement, the two main influencing factors were believed by the IEA to be: policies to support long-term transition, and '…short-term macroeconomic and market trends, which may accelerate—or impede—the transition towards a lower carbon energy future…'. Though the market trends referred to are difficult to predict, it is clear that the same sorts of effects have been present in previous energy transitions. Thus it seems important in considering the transition from hydrocarbons to renewables to consider previous transitions, and also to be aware of the similarities of these processes to those seen in climate change.

TELEOLOGICAL FEEDBACK

The cycles of feedback that are intrinsic to Earth biogeochemical processes and to human systems contain regulatory and destabilising elements. When environmental conditions visibly worsen, efforts are often made to improve things, and human interventions to make improvements in the environment are not a new phenomenon. As I described in an earlier chapter, efforts were made in Pittsburgh in the 1880s to increase the use of gas in steel and iron manufacture to reduce air pollution.

A similar story comes from London in the 1950s. The Great Smog of London between December 1952 and March 1953 was caused by cold air, fog, the diesel buses on London's roads, and thousands of domestic coal fires and coal-fired factories. The smog killed 4000 people, and another 8000 died in the weeks and months that followed. The Clean Air Act of 1956 introduced controls on air pollution, for example 'smoke control areas' in some towns and cities where only smokeless fuels could be burnt. It also encouraged power stations to be sited away from cities, and for the height of some chimneys to be increased. The effects of the Clean Air Act were almost immediate, with levels of SO_2 and smoke falling throughout the 1960s and 1970s.

Better air quality is needed in many modern fast-growing cities. The Chinese government's campaign of 'Defending the Blue Sky' began before the Summer 2008 Olympic Games to improve the poor quality of Beijing's air. This involved relocating more than a thousand heavy industrial and power-generation plants outside the city and the introduction of natural gas for some domestic and industrial use, as well as cleaner coal with reduced sulphur. This had an almost immediate effect on Beijing's atmosphere.

The problem of acid rain is similar. This is caused by the sulphur in coal being oxidised to sulphur dioxide during burning. The dioxides SO_2 and SO_3 dissolve in rainwater, making it acidic and affecting plants and watercourses. Most modern coal power stations are fitted with flue gas desulphurisation (FGD), which removes the SO_2 from exhaust (flue) gas.

The ozone hole above the Antarctic was first reported in a scientific paper in 1985 in the journal *Nature* by scientists from the British Antarctic Survey, though it may have existed as far back as 1976. The hole (and other smaller ozone holes in other locations) were generated due to reactions in the stratosphere between ozone and halogens, mainly from photodissociation of man-made refrigerants, solvents, propellants, and foam agents such as chlorofluorocarbon (CFC), freon and halon.

Action following the discovery of ozone depletion was swift. The Montreal Protocol on 'Substances that Deplete the Ozone Layer', agreed upon in September 1987, is an international treaty designed to protect the ozone layer by phasing out the production of substances that are responsible for ozone depletion. It entered into force on the first of January 1989. The Montreal Protocol has been described by many as a great success, with around 98% of ozone-depleting substances having been phased out. The ozone holes are smaller than they were, although it may take more than 30 years to return them to a more normal form. The Montreal Protocol also had unprecedented support with signatories from 196 states and the European Union.

The interventions I have described, though undoubtedly successful, were made to fix problems that were not necessarily related directly to wider climate change and to the physical and chemical cycles that govern the climate. The idea that is embodied by the philosophy of the Anthropocene is that humankind is capable of changing the environment on a planetary scale, but also that humankind is also capable of making conscious interventions at a scale that can reverse harmful change. This has, of course, begun already. The technology of carbon capture and storage outlined earlier in this book has no other function but to reduce CO_2 in the atmosphere.

But perhaps the main point to make here is that interventions on a planetary scale mark an important change in human and Earth history, where essentially large-scale, conscious, purposeful feedback (also known as teleological feedback) has taken its place amongst the range of other physical and chemical feedbacks, and human and economic trends and pressures.

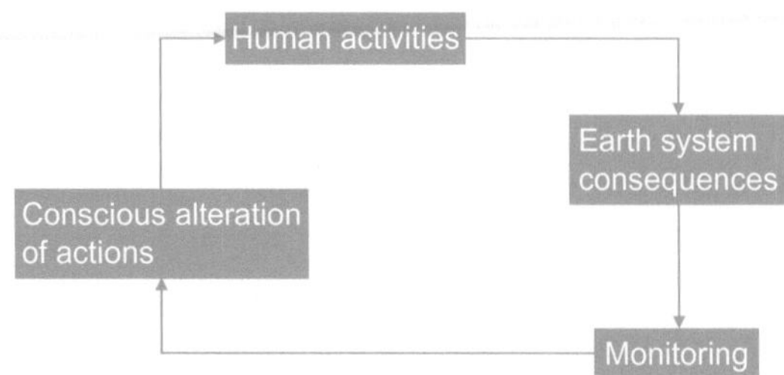

Fig. 7.4 Teleological feedback in the Earth system. Human activities produce changes in the Earth system, these are detected and measured, and conscious changes to activities are then made. *(After Lenton, T., 2016. Earth System Science: A Very Short Introduction. Oxford University Press, 176pp.)*

The feedback consists essentially of four elements (Fig. 7.4), where human activities produce changes in the Earth system, where these are detected and measured, and where conscious changes to activities are made.

Earlier in this book I described the point at which humankind began to be less limited by natural resources when coal was taken up in the British industrial revolution. The historian Andreas Malm sees this commencement of the 'fossil economy' as a seminal event in world history. Perhaps the implementation of the teleological feedback marks a similarly important historical event.

But before we celebrate the establishment of this feedback loop we need to realise that operating it is, or will be, very difficult. The three elements that are particularly needed are deep understanding of the Earth system, sophisticated and sensitive measuring and monitoring—and of course the right kind of intervention. The intervention could be deliberate amplification of negative regulatory feedbacks, for example CCS; or amplification of positive feedbacks to move the Earth through a tipping point from one environmental state into another more desirable (or sustainable) state.

Discussion of this type brings to mind terms such as 'geoengineering' or 'climate engineering'—essentially artificial regulatory negative feedbacks. There are a whole host of ideas, some more practical than others. Geoengineering is broadly of two types: management of the amount of solar radiation reaching the ground, and removal of carbon dioxide from the atmosphere.

In the first, solar radiation management techniques aim to diminish sunlight absorbed by the atmosphere and ground by reflecting it away, or by increasing the albedo of the atmosphere or the Earth's surface. These techniques would not reduce greenhouse gas concentrations in the atmosphere, and so would not (for example) reduce ocean acidification. Some proposed methods include installing reflective roofs on buildings or increasing the cultivation of crops that give the ground a high albedo. Others include increasing the reflectivity of clouds using sea spray, and spreading reflective aerosols in the atmosphere.

In the second category greenhouse gases, mainly CO_2, are removed from the atmosphere via an agent (like newly planted trees) or by chemical or physical means. Examples include the manufacture of inert stable carbon known as biochar, which can be mixed with soil; others include afforestation and reforestation, and ocean fertilisation to increase carbon drawdown. Some would regard CCS and bioenergy with carbon capture and storage (BECCS) as geoengineering of this type.

Geoengineering is sometimes regarded with skepticism by the public and the scientific community. Amongst concerns are the cost of interventions, the unintended consequences or side effects, and the false sense of security associated with 'quick-fixes', rather than considered solutions.

SUMMARY

What I have tried to show in this book so far is that geological science and the materials and processes of the Earth are important to energy and climate change. This chapter shows that human energy systems and climate change are not only linked through the carbon cycle, but also in the sense that they are both systems that are affected by feedbacks and tipping points. These need to be understood in detail, to make the conscious interventions we need to make to avoid dangerous climate change.

The understanding of the Earth system needed to make sensible interventions will consist of a large amount of geological science. The observations and monitoring that humankind will have to do to understand change will have to include the subsurface as much as the surface.

BIBLIOGRAPHY

Bains, S., et al., 2000. Termination of global warmth at the Palaeocene/Eocene boundary through productivity feedback. Nature 407, 171–174.

Farman, J.C., et al., 1985. Large losses of total ozone in Antarctica reveal seasonal ClOx/NOx interaction. Nature 315, 207–210.

Landschützer, P., et al., 2015. The reinvigoration of the Southern Ocean carbon sink. Science 349, 1221–1224.

Lenton, T., 2016. Earth System Science: A Very Short Introduction. Oxford University Press. 176pp.

Pearson, P.J.G., Foxon, T.J., 2012. A low carbon industrial revolution? Insights and challenges from past technological and economic transformations. Energy Policy 50, 117–127.

Rowland, F.S., 2006. Stratospheric ozone depletion. Phil. Trans. R. Soc. B 361, 769–790.

Sluijs, A., et al., 2007. The Palaeocene-Eocene thermal maximum super greenhouse: biotic and geochemical signatures, age models and mechanisms of global change. In: Williams, M., Haywood, A.M., Gregory, F.J., Schmidt, D.N. (Eds.), Deep time perspectives on Climate Change: Marrying the Signal from Computer Models and Biological Proxies. The Geological Society, London, pp. 323–349.

Waters, C.N., et al., 2016. The Anthropocene is functionally and stratigraphically distinct from the Holocene. Science 351, 137–147.

CHAPTER 8

The Geological Macroscope

Contents

In the early 2000s ecologists began to monitor the environment in a more comprehensive way than ever before, using sensors and measuring devices, with the result that they began to understand ecology better than they had before. Meteorologists and oceanographers have been monitoring and measuring for considerably longer, harnessing the power of new sensors, telemetry, and computing to build models to understand complex systems better, often leading to a better ability to predict change and quantify uncertainty. Geologists have been slower to take up this technology, except in the fields of volcanology and seismology. In this chapter I describe what geologists will have to do to develop a geological 'macroscope' which allows large objects to be studied with numerous sensors, and the benefits that will accrue, including a fuller understanding of the Earth's environment (including the subsurface), better ability to manage the subsurface to keep our activities sustainable, and an ability to plan the interventions to prevent dangerous environmental change.

To operate a successful teleological feedback system and of course to understand our complicated globe and the many stresses and dependencies within it, we need to be able to monitor and observe the environment in detail, and collect data. After collecting these data, we will have to build mathematical or computer models that encapsulate our understanding of the systems. We can use this understanding to forecast and predict change, and to plan our interventions.

Energy and Climate Change
https://doi.org/10.1016/B978-0-12-812021-7.00008-7

By 'environment' I mean not just the atmosphere and the surface parts of the water cycle, but also the subsurface. Often the subsurface has not been seen as part of the environment. In the past it has not been easy to develop understanding of the subsurface because the range of sensors was limited. For these reasons geological science has rather lagged behind other sciences in the sophistication of its monitoring, except perhaps in seismology and volcanology. However, sensors are becoming cheaper, more complex, and more robust and can be deployed at the surface and the subsurface. The computing power to deal with data telemetered from sensors is already here, and the ability to image and visualise the subsurface is getting better.

Meteorology and oceanography in particular have enormously gained from sensor and telemetry technology and increases in computer modelling capacity and visualisation. Sciences like ecology are just beginning to benefit.

MACROSCOPE IN THE REDWOODS

An example from ecology shows some of the scientific gains from systematic monitoring and measuring using sensors. This is a well-known study described in a scientific paper published in 2005 called 'A macroscope in the redwoods'. The paper has been cited nearly a thousand times in the peer-reviewed scientific literature, but many more times in the scientific media.

In the paper, the authors introduce the idea of the 'macroscope', which is essentially a way of looking at something very large in great detail using a large numbers of sensors, or as the authors of the paper say:

> *'Wireless sensor networks offer the potential to dramatically advance several scientific fields by providing a new kind of instrument with which to perceive the natural world. As the telescope allowed us to perceive what is far away and the microscope what is very small, some refer to sensor networks as "macroscopes" because the dense temporal and spatial monitoring of large volumes that they provide offers a way to perceive complex interactions'.*
>
> ***(Tolle et al., 2005)***

In the case of this particular scientific study, the large 'volume' that was studied was a 70-m tall redwood tree (Fig. 8.1). A wireless sensor network installed by the scientists recorded 44 days in the life of the tree continuously, taking measurements every 5 min in a dense three-dimensional array. Each sensor

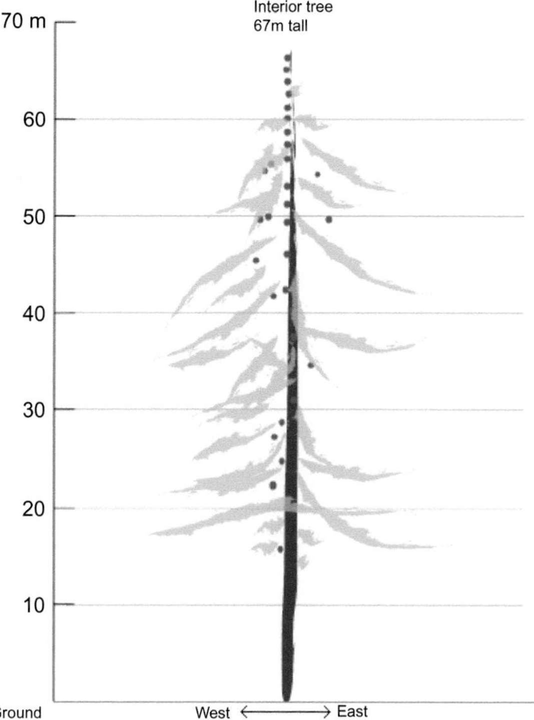

Fig. 8.1 Approximate positions of sensors in a redwood tree. A wireless sensor network in a dense three-dimensional array took measurements every 5 min. *(From Tolle, G. et al., 2005. A Macroscope in the Redwoods. In: Redi, J.K. (Ed.), SenSys'05 Proceedings of the Third International Conference on Embedded Networked Sensor Systems. San Diego, California, USA, vol. 2, Association for Computing Machinery, New York, pp. 51–63.)*

measured air temperature, relative humidity, and photosynthetically active solar radiation (PAR); and following computer processing, the network captured a detailed picture of the ecology and microclimate of the tree. Trends in time and space that had not been seen before were revealed by the study, and were illustrated very clearly in a series of charts and diagrams. One of these is reproduced in Fig. 8.2.

The data indicate trends over a 24-h period as well as the same variables at different levels in the tree. One way in which these data were used was to investigate the rate of sap flow in response to humidity, air temperature, and PAR. Scaling up such work could allow botanists to understand part of the large-scale processes of carbon and water exchange within a forest ecosystem. In the way that ecology has benefitted from this and subsequent

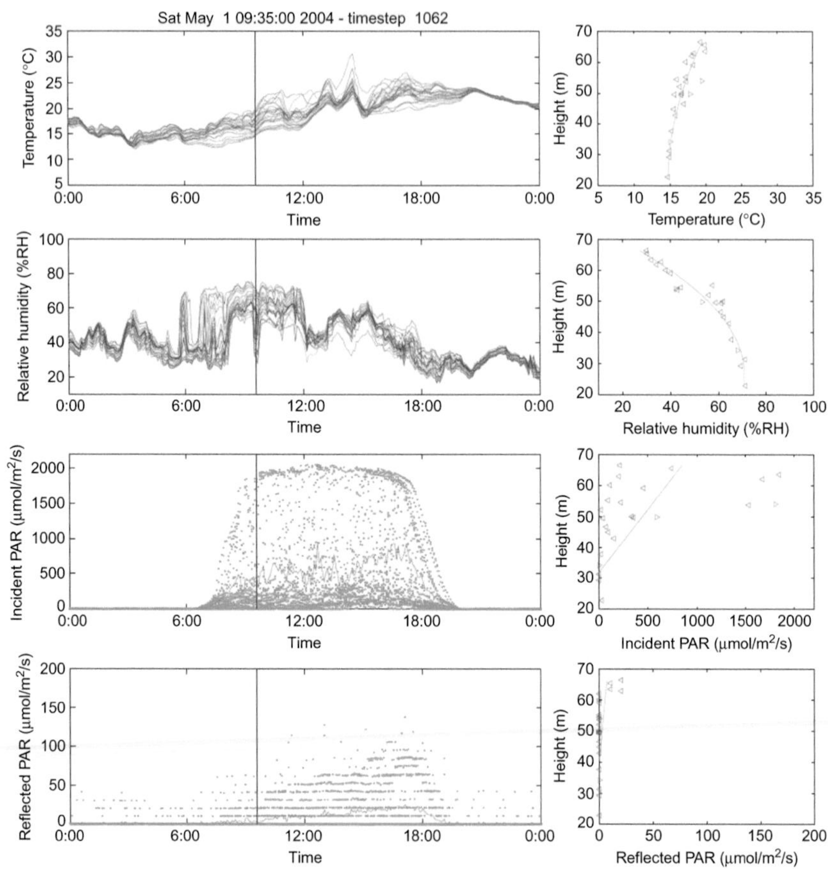

Fig. 8.2 Trends in air temperature, relative humidity, and photosynthetically active solar radiation (incident and reflected) over a 24-h period, as well as the same variables at different levels in the tree. *(From Tolle, G. et al., 2005. A Macroscope in the Redwoods. In: Redi, J.K. (Ed.), SenSys'05 Proceedings of the Third International Conference on Embedded Networked Sensor Systems. San Diego, California, USA, vol. 2, Association for Computing Machinery, New York, pp. 51–63.)*

work, the application of similar 'macroscopes', this time installed in the subsurface, would provide advances in geological science.

MONITORING IN METEOROLOGY AND OCEANOGRAPHY

Meteorology and oceanography were the first of the environmental sciences to collect and process large amounts of data from sensors.

Most weather stations provide measurements of temperature, pressure, wind, and humidity, but more sophisticated stations have sensors to detect dust, smoke, methane, ozone, carbon monoxide, and carbon dioxide. Sensors are also capable of measuring the height of the cloud ceiling, lightning, infrared and ultraviolet light, radiation, and visibility. Radiosondes are launched to sense the higher atmosphere. Civil and scientific aircraft, satellites, and drones collect meteorological data, radar, and LIDAR (Light Detection And Ranging). With data of this type, agencies such as the UK's Meteorological Office produce forecasts using software known as the Unified Model run on one of the world's most power computers. A 36-h forecast for weather is produced for the UK and surroundings, a 48-h forecast for Europe and the North Atlantic, and a 144-h forecast for the globe.

Ocean monitoring is well established (Fig. 8.3). Ocean-scanning satellites map ocean-surface topography caused by ocean currents, and ocean warming and cooling. Other satellite instruments measure the direction and magnitude of the effect of wind on the sea surface, surface water temperature, the distribution of chlorophyll, and precipitation over the ocean. Acoustic sensors can supply 3D images of temperature variability from the surface down to mid-depth, and long-term changes in temperature at depth. Ocean research vessels and drifting and anchored buoys measure temperature, salinity, and currents in the upper water layers. Tide gauges measure variations in monthly and shorter-period mean sea level. These measurements and observations contribute to programs such as the World Ocean Circulation Experiment (WOCE) and the Tropical Oceans and Global Atmosphere (TOGA) programme. WOCE and TOGA are designed to understand the changing oceans, for example variations in the Indian Ocean monsoon and droughts, connections between oceanic and atmospheric processes, and the ocean carbon cycle.

MONITORING IN SEISMOLOGY AND VOLCANOLOGY

Geological sensor systems that are well advanced include those of seismology for earthquakes, and wider subsurface monitoring for volcanoes.

An earthquake warning system uses sensors such as accelerometers and seismometers which feed into computers, which can if needed issue alarms to warn people if an earthquake is starting. Though the systems are similar, in the extent of data gathered, to those of meteorology or oceanography,

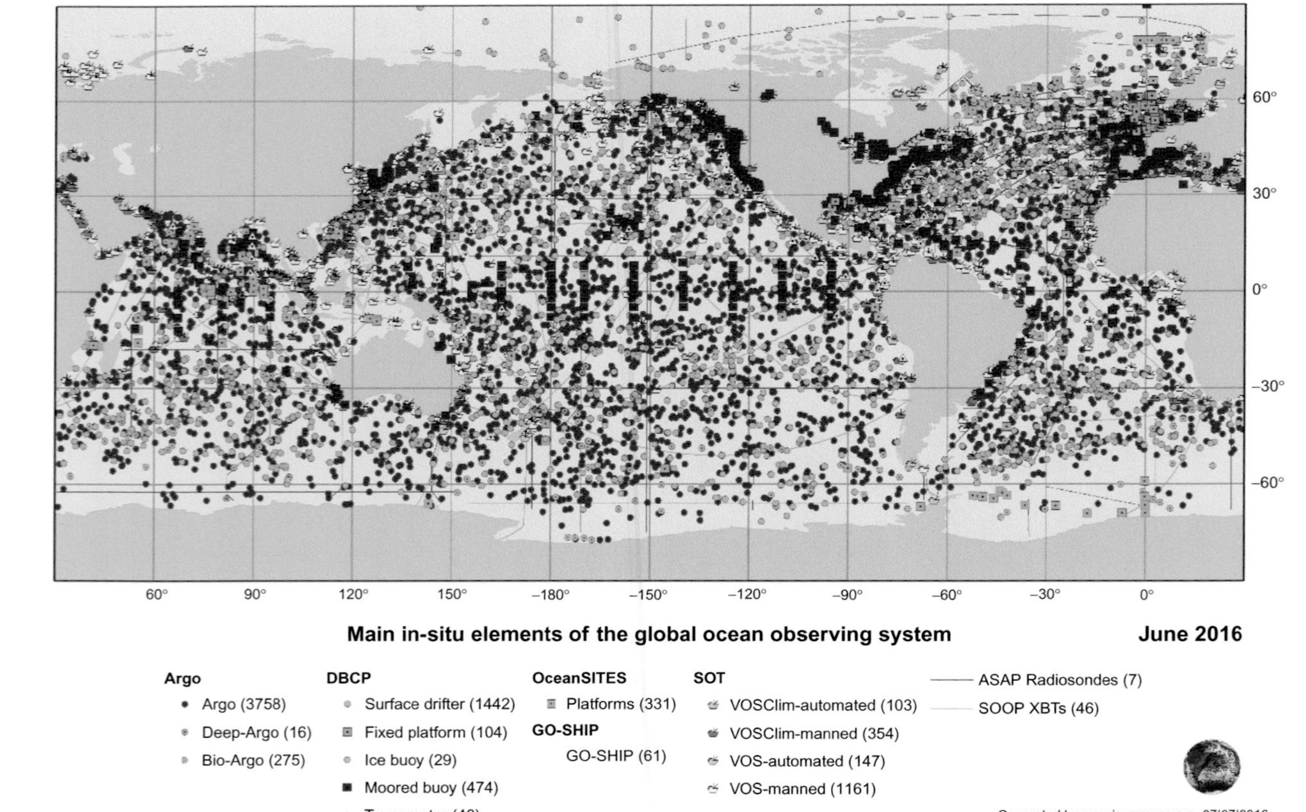

Fig. 8.3 Sensors and measuring devices of the Global Ocean Observing System. *(From GOOS, JCOMM, IOC/UNESCO. http://www.goosocean.org/index.php?option=com_content&view=article&id=24&Itemid=123.)*

earthquake warning systems are more regional. These systems also do not offer 'forecasts' in the sense of being able to warn decisively of future earthquakes.

Volcano monitoring involves seismology mainly because increasing seismic activity is a good indicator of increasing eruption risk. Gas emissions are also monitored because, as magma nears the surface and its pressure decreases, gases escape. The bulging of a volcano signals that magma has accumulated near the surface, and this can be detected by precise measurements of the tilt of a slope. The way that a volcano heats up can be sensed through infrared band satellite imagery and the measurement of associated hot springs and fumaroles. Like earthquake warning systems, volcano monitoring is necessarily regional and local and is focused on warnings and short time-scale effects on human populations.

Perhaps the most ambitious of attempts to collect data on earthquakes and tectonic plate movements is EarthScope, a large science programme funded by the US National Science Foundation and operated in collaboration with the United States Geological Survey (USGS) and the National Aeronautics and Space Administration (NASA). EarthScope scientists use seismometers, very precise GPS measurements of the Earth's magnetic field, and images from aircraft and satellites to describe and model, and thereby develop better warnings of natural hazards. EarthScope operates a very large array of seismometers that will be deployed in the years to come across the whole of the continental United States. EarthScope also operates the Plate Boundary Observatory (PBO), which measures the strain or distortion across the boundary between the Pacific and North American tectonic plates in the western United States by measuring precise movements with sensitive GPS, borehole strainmeters, and tiltmeters. Perhaps most exciting, for geologists at least, is the San Andreas Fault Observatory at Depth (SAFOD) facility, which is centred around a 3.1-km deep borehole drilled directly into the San Andreas Fault midway between San Francisco and Los Angeles. The fault at this location has moved six times since 1857, and so the borehole observatory could provide the ability to see earthquakes actually starting.

A similar initiative in Europe is the European Plate Observing System (EPOS), which aims mainly to bring together and harmonise geophysical data from Europe's observatories that monitor earthquakes, volcanoes, and the strain and distortion of rocks.

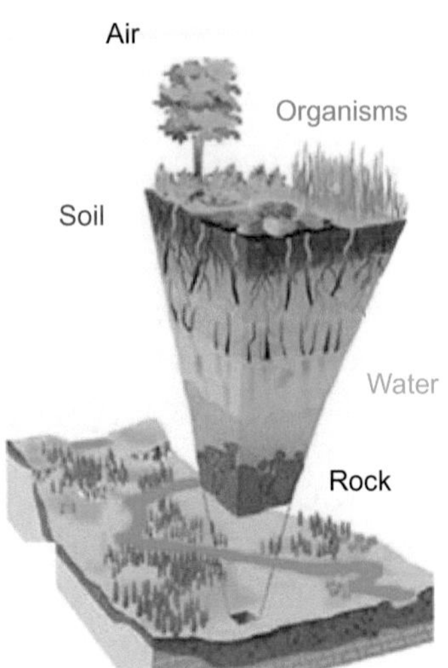

Fig. 8.4 The elements of a CZO. *(The critical zone. Illustration modified from Chorover, J., et al. 2007. Soil biogeochemical processes in the critical zone. Elements 3, 321–326.)*

CRITICAL ZONE OBSERVATORIES

Critical zone observatories (CZOs) monitor the zone between the top of the vegetation cover and the base of the weathered zone in rocks, and so have a different aim to initiatives like Earthscope and EPOS. This interface between the surface and subsurface is considered critical because it is the zone of freshwater, soils, and terrestrial life (Fig. 8.4), and therefore provides most of the ecosystem services on which humankind depends. In the United States CZOs have been set up in areas such as Boulder Creek, Colorado; Calhoun, South Carolina; Eel River, California; the Susquehanna Shale Hills in Pennsylvania; and Southern Sierra. CZOs tend to study natural and unperturbed environments aiming at understanding environmental baseline conditions, though a few, such as the Susquehanna Shale Hills CZO, look at the effects on the natural environment of hydrocarbon exploitation.

CITIZEN SCIENCE AND THE 'INTERNET OF THINGS'

A big change in the way that sensed data are being collected began in the last few years. This is the realisation amongst scientists that data could be usefully collected by people that are not an intrinsic part of the scientific process. This is known as 'citizen science'. Similarly, objects (like household appliances) primarily for other uses can be used to collect data. This is part of the concept of the 'internet of things'.

Citizen science has a relatively long history, but the idea that technological devices that people carry, for example mobile phones, can be used to collect large amounts of continuous data is relatively new. A good example is the iShake app (Fig. 8.5).

iShake uses the accelerometer inside an iPhone to collect earthquake data and create intensity maps of earthquakes for scientific research and potentially, in a large earthquake, to help in rescue efforts. The system can also work to provide a phone user with real-time broadcasts of data, for example warnings and advice. The British Geological Survey's mySoil and myVolcano apps allow data on soil and erupting volcanoes to be submitted for research. The Marine Debris Tracker app, a joint partnership of the National Oceanic and Atmospheric Administration and the University of Georgia, allows citizens to submit information about marine debris.

Fig. 8.5 The iShake mobile phone app. *(From http://citris-uc.org/contact/.)*

The internet of things is a network of physical 'smart devices' including buildings, vehicles, and other items containing electronics, software, sensors, and network connectivity. These can collect and exchange data and be controlled remotely across networks, so data about the physical world can be recorded in computer-based systems without human intervention. In practice this means, for example, that smart electric power grids can manage themselves to adjust to power demand; similarly, 'smart homes' can manage power use better. Experts estimate that the internet of things will consist of about 30 billion objects by 2020.

REASONS TO MEASURE AND MONITOR THE SUBSURFACE BEYOND EARTHQUAKES AND VOLCANOES

Beyond the need to observe, monitor, and model the subsurface as part of the Earth system to improve our ability to operate teleological feedback, are there any more pressing needs to monitor the subsurface beyond earthquakes and volcanoes? There are several aspects of subsurface usage that have been covered in earlier chapters in this book that use sophisticating monitoring, for example for the management of subsurface gas storage in salt caverns or depleted gas or oil fields. Oil and gas companies also routinely monitor the integrity of their reservoirs and the way that fluids move in the reservoirs. An example is the Groningen field in the Netherlands, which started producing natural gas in 1963 and is expected to end production around 2080. At the start of production, reservoir pressure was around 350 bar, but by 2016 it had dropped to about 100 bar because much of the gas had been extracted. This resulted in the sandstone gas reservoir becoming more compacted, and a series of earthquakes related to subsurface collapse. The reservoir and rocks above it (known as the overburden) are therefore monitored, and extraction and injection carefully managed to minimise seismicity.

However, one of the areas where monitoring is perhaps currently inadequate is fugitive emissions. As I explained in an earlier chapter, an argument against the take-up of shale gas as a substitute for other fossil fuels is that, at times, it has been associated with leaks of gas into the atmosphere, either through surface pipework, subsurface well infrastructure, or from surface open flowback ponds. This kind of leak is particularly a problem because methane is a potent greenhouse gas, around 25 times more powerful than carbon dioxide over a 100-year timescale.

In fact, fugitive emissions are being seen more as a problem of the oil and gas business and industry in general, rather than just shale gas. Studies in 2000 of greenhouse gas emissions from well, pipework, and refining in Canada estimated that associated fugitive emissions have a global warming potential equal to 17 million tonnes of carbon dioxide. Activities like deliberate venting of methane and flaring add more greenhouse gases to the atmosphere. Above-ground monitoring is carried out in refineries and at production facilities, but there are grounds to think that these underestimate the full scale of leakage. A new technology, differential absorption LIDAR (DIAL), has shown this through imaging and measurement from a location far enough away to gain a full appreciation of a particular site's emissions. A recent DIAL survey suggested that emissions at a Canadian refinery were much higher than those previously reported using conventional monitoring methods.

This kind of surface monitoring may be missing emissions of gases that do not find their way as far as the refinery or even to the top of the well. Those emissions that penetrate faulty well infrastructure at depth may find their way into deep rock layers and travel upward into shallow rock layers and aquifers. How do we know how much 'stray gas' finds its way into the subsurface and then eventually into the soil, rivers, groundwater, and finally into the atmosphere? At the moment, little is known of this possibly large flux. Of course, flows of hydrocarbon gases through rock layers are also natural because gas will tend to find its way to the surface in many ways (for example the eternal flames of Kirkuk in Iraq featured in an earlier chapter). The difficulty is to distinguish natural and 'anthropogenic' flow, though of course for understanding biogeochemical cycles and their effects, any flow is important.

One way to distinguish natural and anthropogenic flows is to use geochemical techniques such as ratios of compounds or elements associated with a particular fluid, and the isotopic composition of that fluid. Taking again methane, the carbon of biogenic methane generated by microorganisms in aquifers has a lower $\delta^{13}C$ than that of thermogenic methane (essentially conventional or unconventional hydrocarbon gas). This difference was used very effectively recently by a team led by Lisa Molofsky studying 'stray' methane in shale gas fields in Pennsylvania. Molofsky was able to distinguish biogenic and thermogenic methane, and also the slightly different thermogenic methane types from different shale layers (Fig. 8.6).

This crucial finding seems to suggest that in this case stray gas—though clearly thermogenic—probably leaked from shale layers above the

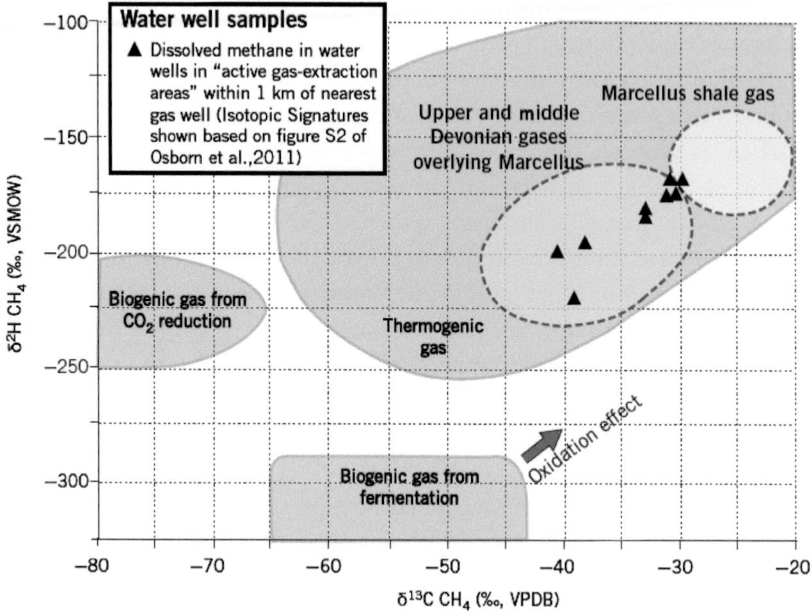

Fig. 8.6 A scatter graph used to distinguish different kinds of methane by $\delta^{13}C$ and hydrogen isotope ratios in methane. The methane samples in the scatter graph (black triangles) appear in an area indicating shale layers above the Marcellus shale. *(From Molofsky, L.J. et al., 2011. Methane in Pennsylvania water wells unrelated to Marcellus shale fracturing. Oil Gas J. 54–67.)*

hydraulically fractured layer (in this case the Marcellus shale), suggesting that hydraulic fracturing was not the direct cause of stray gas, but that leaking or faulty production wells were.

Problems with leaking wells do not appear to be an isolated problem. According to recent studies by a team led by Richard Davies of Newcastle University, amongst 8030 wells drilled into the Marcellus shale between 2005 and 2013, 6.3% were reported for infringements related to well integrity failure. Even in the UK, where the onshore oil and gas business is relatively small, there is evidence that old oil and gas wells are hard to find and that the ownership of these old lost wells is often unknown. In the United States and Canada, where the ground is penetrated by literally millions of oil and gas wells, the need for monitoring is vital. In Alberta, Canada, for example, there are believed to be around 316,500 wells. Drilling started in Alberta late in the 19th century, but drilling and production were not regulated until the late 1930s. This means that many wells may be of poor quality. Many wells are probably 'lost'.

Methane does not only leak from abandoned or lost oil and gas wells, but also from abandoned and working coal mines. Globally, coal mining is believed to be responsible for 8% of total anthropogenic methane emissions.

ENVIRONMENTAL BASELINE SURVEYS

To identify stray methane, some idea of the natural levels of methane in groundwater or in rock needs to be established before changes can be confidently attributed to human activities such as oil and gas extraction or leaking wells. The British Geological Survey are conducting baseline measurements in several parts of the UK. Early work showed that natural concentrations of methane ranged from <0.05–42.9 μg/L (micrograms per litre) for groundwater in Chalk, the aquifer that supplies much water for London; and <0.05–465 μg/L for Permo-Triassic sandstone, which supplies groundwater in the English midlands. Incidentally these figures imply emissions to the atmosphere following extraction of groundwater which are 0.05% of all UK methane emissions. More detailed recent baseline measurements on groundwater and seismicity in the northwest and northeast of England in advance of possible production of shale gas measure a range of groundwater variables (including pH, electrical conductance, total dissolved gas, water level, and water temperature) continuously or almost continuously. The data are also displayed on the internet in real time or near real time (Fig. 8.7).

Baseline surveys can also use other techniques such as electrical resistivity tomography, which measures the electrical resistivity of the ground and can therefore detect fluids and how they move, for example to show the interaction of salt water and fresh water in coastal aquifers.

An example is an electrical resistivity imaging system installed in the normally dry bed of the River Andarax in Almeria, Spain to monitor the seawater-freshwater interface in an underlying shallow aquifer. A buried electrode array extends inland 1.6 km upstream from the shoreline (Fig. 8.8). The unmanned system generates a daily time-lapse image to help understand the way that the seawater-freshwater interface changes naturally. Analysis like this can help regulate pumping and irrigation schemes. Similar systems could be installed to detect deeper flows and show the natural rhythms of subsurface processes as part of the baseline.

Monitoring boreholes could also be fitted with sensitive instruments that measure movement like tilt or distortion. Satellites using interferometric

Home » Our research » Groundwater » Shale gas » Environmental monitoring » Vale of Pickering » Real-time monitoring data

Environmental baseline monitoring in the Vale of Pickering—Real-time monitoring data

Monitoring
Air composition
Groundwater quality
Seismicity

BGS data overlay
(none)

Base map
Road map

reset zoom

Newton on Rawcliffe
Cropton
Carlton
Kirkbymoorside
Wrelton
Nawton
Helmsley
Great Edstone
Pickering
Duncombe Park
A170
Kirby Misperton
VALE OF P
Newburgh Priory
Gilling Castle
Hovingham
Ryton
Low D
Leaflet | Powered by Esri | DoBH, OS, Esri, HERE, Garmin, NGA, USGS

pH
interactive graph*

Specific electrical conductance
interactive graph*

Total dissolved gas
interactive graph*

Water level
interactive graph*

Water temperature
interactive graph*

Fig. 8.7 Real-time or near real-time environmental monitoring data on groundwater in part of northern England. The map shows the location of boreholes which contain sensors, and the graphs following show some of the variables measured. *(From the BGS website: http://www.bgs.ac.uk/research/groundwater/shaleGas/monitoring/vopDataSummary.html?. BGS © NERC [2017]. Contains Ordnance Survey data © Crown Copyright and database rights [2017].)*

Fig. 8.8 An electrical resistivity imaging system installed in the bed of the River Andarax in Almeria, Spain. Left is the array at the surface. Right shows an image of the seawater-fresh water interface in an underlying shallow aquifer. *(From http://www.bgs.ac.uk/research/tomography/ALERT.html, BGS © NERC [2017].)*

synthetic aperture radar (InSAR) can look for tiny amounts (millimetres) of ground deformation or subsidence. Satellites can, in the right circumstances, even look for leaking gases at the surface.

SUMMARY

This chapter discussed how the changing climate and new developments in energy, storage, and land use will mean that geologists will need to be able to understand subsurface processes better. A technological revolution in sensors, imaging, and computing is going on around us. Early adopters of this technology in environmental science have been meteorologists and oceanographers, but now other environmental scientists and ecologists are taking up the technology. But large-scale monitoring and measuring in real time has not been a feature of geological science outside volcanology and seismology, and geologists have largely missed out.

The 'geological macroscope' and associated sensor deployment will enable better 3D models for prediction of such events as groundwater flooding, cliff falls, and erosion around our coasts, as well as effects that climate change might have on the landscape or such built infrastructure as railway embankments. It will help us build better models for subsurface developments in cities and rural areas, for example geothermal for heating and air conditioning, gas storage, compressed air energy storage, nuclear waste disposal, and carbon capture and storage.

Geology, unlike many modern sciences, has a strong descriptive and qualitative element where individual judgement and description still count for a lot. The disadvantage of this qualitative approach is that, at times, geologists find it difficult to precisely quantify the uncertainty associated with the observations and interpretations, for example the lines drawn on geological maps. The geological macroscope will take objective measurements and so allow more accurate estimates of uncertainty.

BIBLIOGRAPHY

Bank, J., 2012. Barriers and Opportunities for Reducing Methane Emissions from Coal Mines. In: Clean Air Task Force. 22pp.

Chambers, A., et al., 2008. Direct measurement of fugitive emissions of hydrocarbons from a refinery. J. Air Waste Manage. Assoc. 58, 1047–1056.

Chambers, J.E., et al., 2014. 4D Electrical Resistivity Tomography monitoring of soil moisture dynamics in an operational railway embankment. Near Surface Geophys. 12, 61–72.

Clearstone Engineering 1994. A National Inventory of Greenhouse Gas (GHG), Criteria Air Contaminant (CAC) and Hydrogen Sulphide (H2S) Emissions by the Upstream Oil and Gas Industry, Volume 1, Overview of the GHG Emissions Inventory. Canadian Association of Petroleum Producers: v. Retrieved 2008-12-10.

Davies, R.J., et al., 2014. Oil and gas wells and their integrity: Implications for shale and unconventional resource exploitation. Marine Petroleum Geol. 56, 239–254.

Gooddy, D., Darling, G., 2005. The potential for methane emissions from groundwaters of the UK. Sci. Total Environ. 339, 117–126.

Molofsky, L.J., et al., 2011. Methane in Pennsylvania water wells unrelated to Marcellus shale fracturing. Oil Gas J. 54–67.

Molofsky, L.J., et al., 2013. Evaluation of methane sources in groundwater in northeastern Pennsylvania. Ground Water 51, 333–349.

Revelle, R., Bretherton, F., 1986. Global ocean monitoring for the World Climate Research Programme. Environ Monit Assess (1), 79–90.

Tolle, G., et al., 2005. A macroscope in the redwoods. In: Redi, J.K. (Ed.), SenSys'05 Proceedings of the Third International Conference on Embedded Networked Sensor Systems, vol. 2, San Diego, CA. Association for Computing Machinery, New York, pp. 51–63.

de Waal, J.A., et al., 2015. Production induced subsidence and seismicity in the Groningen gas field—can it be managed? Proc. IAHS 372, 129–139.

Watson, T.L., Bachu, S., 2007. In: Evaluation of the Potential for Gas and CO_2 Leakage along Wellbores. Alberta Energy Resources Conservation Board, Edmonton, AB. Canada SPE Paper #: 106817.

CHAPTER 9

Energy and Climate Change: Geological Controls, Interventions, and Mitigations

Energy and climate change are often rightly linked by policymakers and politicians because, since the industrial revolution, the fossil fuels that have powered the global economy have also altered the atmosphere and are highly likely to cause climate change that may be difficult and expensive to adapt to. In this chapter I summarise the main points of the book and clarify the role that geological science plays in energy and climate change, specifically in controls, interventions, and mitigations.

The geological controls on energy and climate change are clear. The geology of fossil fuels has had a role in governing past industrialisation, human wealth, and economic development. Furthermore the distribution of fossil fuels in the developing world may affect further growth and the ability of the world to keep within the limits of the Paris Accord. Geological interventions like carbon capture and storage may allow large-scale industrial emissions reduction. Bioenergy with CCS (BECCS) could provide net negative emissions, though there are questions over its sustainability. The use of groundwater, particularly in the developing world, will be an important mitigation to survive water stress caused by climate change. But to make these interventions and mitigations work to their best effect, geological science must make use of more comprehensive subsurface monitoring.

In this book I have tried to consider climate change and energy as scientific concepts that are fundamentally linked through time, through the carbon cycle and through other biogeochemical cycles. The first chapter describes how fossil fuels are formed and how, as their name implies, they are intrinsically related to biological processes and to the long-term geological carbon cycle. Through their formation, fossil fuels alter climate by removing or sequestering carbon for the long term in geological 'storage'. Probably relatively little of this storage is 'permanent' through geological timescales, because the carbon that is buried is liable to appear at the surface again—and through natural surface oxidation this newly disinterred carbon will likely release CO_2 back into the atmosphere.

Energy and Climate Change
https://doi.org/10.1016/B978-0-12-812021-7.00009-9

In the second chapter I demonstrated that the huge span of geological time shows many instances of natural climate fluctuations that have an endogenous Earth system origin, such as volcanicity, some of which have been severe enough to affect the continuity of life on Earth, for example at the Permian-Triassic boundary and the Palaeocene-Eocene Thermal Maximum, respectively 252 million and 55 million years ago. These events show that the Earth system does return to 'normality' (depending on how this is viewed), and geological processes working through the carbon cycle are usually responsible, though oceanographic and other processes also have a role.

The third chapter reveals just how important human activities are in relation to climate change, in fact just as influential as natural processes. The problem is that humankind short circuits the long-term geological carbon cycle by burning fossil fuels and upsetting a complex balance or equilibrium. This reinforces the proposition that climate change and energy are linked by geology but also by human use. Perhaps the starkest representation of transformational human activity is the commencement of the industrial revolution in Britain in the late 18th century, where coal was taken up in earnest as a fuel. The resulting 'inflection point' on the CO_2 curve indicates an important point in human history when the focus of energy resource provision switched from the surface of the Earth to the subsurface. The transition from coal to oil generated atmospheric change too. Changes in the 1950s in the rate of human consumption and manufacturing have apparently generated an 'inflection point' known as the Great Acceleration.

Chapter 4 looks at how this Great Acceleration might proceed, given that the developing world is probably poised to industrialise and to experience changes in living standards, wealth, and energy usage. Most forecasts suggest that energy demand will increase in the developing world, but the extent to which this demand will be satisfied by fossil fuels is not known. The 'inertia' that energy systems carry will tend to help them persist, as will the enormous resources of fossil fuels that exist in the developing world. It is clear that for the developing world to 'leapfrog' fossil fuels will require large amounts of investment in renewables, but early assessments suggest that the developing world, for example Africa, has enormous potential for solar and wind.

Chapter 5 describes how geological processes and materials are not only intrinsic to the long-term carbon cycle, and therefore to energy and climate change, but also can be part of the solution to climate change. The bulk of the chapter is taken up by a description of the technology known as carbon capture and storage (CCS) and how it could be a counterbalance to the fossil-fuel short circuit that I described above. In effect, if we combust carbon, we can also collect the CO_2 from that combustion in power stations and factories and bury it in the subsurface before it reaches the atmosphere. The technology is expensive and unproven as a full chain at full scale, but it is also the only current way that many industrial processes such as cement and ammonia manufacture can be decarbonised.

A variant on CCS known as bioenergy and CCS (or BECCS) is a net negative-emissions technology which has significant risks but is seen by some as the last resort if emissions are not cut quickly enough, for example if the developing world takes up fossil fuels.

Chapter 6 looks at how the subsurface could play a part in how humankind adapts to climate change. The main 'vector' of climate change is likely to be water, and groundwater in aquifers provides backup for surface water variability in relation to seasonal variation and perhaps to more long-term climate change. There are still big research questions over how climate change will affect groundwater quality and recharge. It is also difficult to predict where the most intense future development will take place in the developing world, though perhaps present and future 'development corridors' may allow greater understanding of the food-water-energy nexus in the decades to come.

Chapter 7 perhaps shows most clearly the need to understand and study the processes of the past, but including the recent industrial human past as well as geological history. Human energy systems—the economies that are built around coal, oil and gas—contain mechanisms that operate in similar ways to the physical science feedbacks and tipping points of the natural climate system, and many other natural systems and cycles. There are serendipitous events that lead to the increased use of fossil fuels, and positive feedbacks that allow fuels to rapidly grow. When one energy system transitions to another, a tipping point has been passed. What can we learn from this connection? How can the transitions between energy systems be better managed? Can they be managed?

What I think is the most important message of this book comes in Chapter 8. Technology, as well as having lifted living standards and health, has also placed humankind at odds with its environment, perhaps beginning with the adoption of the fossil economy; but it also delivers the means to help us to adapt better through helping us monitor, measure, and understand the environment. The ability to intervene in an intelligent way to reduce climate change, or better adapt, can only come from a greater understanding of Earth processes. From surface environmental monitoring comes data to help operate the teleological feedback, but geological subsurface data are needed too because aspects of the processes we need to understand operate there.

The title of this book mentions geological controls, interventions, and mitigations. The geological controls on energy and climate change are very strong. The geology of fossil fuels, their extent, distribution, and accessibility have governed in the past the degree of industrialisation, and to some extent the accumulation of human wealth and power. The nations of the industrial revolution are still amongst the most powerful in the world. The extent, distribution, and accessibility of fossil fuels in the developing world may have an effect on how these countries continue to develop and of course on the ability of the world to keep within the limits of the Paris Accord. For the developing world to take another course will require an enormous effort of investment—and political will—to break the inertia of fossil fuels, in effect to take the energy system across a tipping point into a different system.

Geological interventions use the link of the carbon cycle between energy and climate change to bring about useful change. The most obvious of these geological interventions is carbon capture and storage, which seeks to short circuit the carbon cycle, burying carbon faster than the natural system. The variant bioenergy with CCS (BECCS) could provide a strong intervention in the form of net negative emissions, though it is associated with uncertain sustainability.

Finally, mitigation, defined in dictionaries as 'the action of reducing the severity, seriousness, or painfulness of something', has a geological aspect, because geology is capable of making the effects of climate change less severe. The best example is groundwater which, particularly in the developing world, provides humankind with a way to survive water stress.

Geological science is part of a family of sciences that describe crucial parts of the environment and allow it to be monitored and understood—in order to plan the interventions and mitigations that we might want to make to keep this world habitable. But geological science also needs to take up the enormous opportunity of new sensors and computing to work alongside meteorology, ecology, oceanography, and other environmental science to manage the environment of the planet to the benefit of all its inhabitants.

GLOSSARY

Albedo a measure of how much light that hits a surface is reflected without being absorbed

Anoxia in geology and environmental science, the condition in which environments contain very little or no oxygen

Anthropocene a proposed epoch of geological time in which human activity has been the dominant influence on climate and the environment

Aquifer a rock layer containing groundwater

Baseload electricity the core electricity supply

Biogenic formed by biological processes

Bituminous coal soft coal containing a bitumen which is of higher quality than lignite but of lower quality than anthracite

Carboniferous period a period of time in Earth's history between about 300 and 360 million years ago

Casing the lining of a well that prevents fluids passing between the well and the rock it passes through, and that prevents the well from caving in

Coalification chemical and physical transformation of vegetation into coal

Conventional hydrocarbons oil and gas extracted from high permeability rocks, usually from single discrete geological structures

Cracking the process of breaking down large molecule hydrocarbons into simpler smaller ones

Earth's mantle the mainly solid bulk of the Earth's interior lying between the dense core and the thin outer crust

El Niño Southern Oscillation (ENSO) a cyclical variation in winds and sea surface temperatures over the tropical eastern Pacific Ocean which affects the weather patterns of the tropics and subtropics

European Union Emissions Trading Scheme a 'cap and trade' trading scheme aimed at reducing greenhouse gas emissions

Eutrophication excessive amount of nutrients in a lake or other body of water

Fault a crack or fracture in the Earth along which movement can occur or has occurred

Flaring controlled burning of waste gases

Flowback the waste fluid that returns to the well and the surface after hydraulic fracturing

Frack fluid the fluid used in hydraulic fracturing

Fracking also known as hydraulic fracturing. The fracturing of deep rock using high-pressure fluid

Geological Disposal Facility an underground store containing high-level nuclear waste

Geothermal heat from rocks

Greenhouse gas a gas that in the atmosphere allows heat to accumulate, causing the greenhouse effect

Groundwater water naturally distributed in rocks underground

Horizontal well a well drilled vertically (at first) and then turned to horizontal deep down to follow a shale or other rock layer

Hydrogen economy a large-scale industrial system to deliver energy using hydrogen

Immature in petroleum geology, where a rock containing organic matter has not been heated enough to produce oil and/or gas

Liquefied natural gas natural gas (mainly methane) that has been converted to liquid form for ease of storage or transport

Load following a load-following power plant adjusts its power output as demand for electricity fluctuates throughout the day

Mature in petroleum geology, where a rock containing organic matter has been heated enough to produce oil and/or gas

Methane a fossil fuel with the chemical formula CH_4, which is the most common component of natural gas

Migration (in oil and gas) movement of hydrocarbons out of a source rock

Mountain building all the geological processes, chiefly plate tectonic movements, that cause mountains to be formed

Natural gas a fossil fuel, usually methane, from rocks

New Policies Scenario an energy forecast scenario of the IEA which takes account of broad policy commitments and plans that have been announced by countries and their governments, including national pledges to reduce greenhouse-gas emissions and plans to phase out fossil-energy subsidies

North Atlantic Drift a warm ocean current, the eastern extension of the Gulf Stream

Peer review the evaluation of work of one or more scientists by other scientists of similar standing

Permeability the ability of a rock to allow fluids to flow through it

Porosity the amount of pore space or void between the constituent particles of a rock

Produced water water from the deep subsurface produced as a by-product of oil and gas drilling

Reserve the proportion of an oil and gas resource that might be possible to extract given economic and environmental limits

Reservoir rock a rock capable of containing fluids mainly in the pore spaces between its particles

Resource in petroleum geology this is the amount of oil and/or gas in the rocks. This is distinct from the amount that might be possible to extract given economic and environmental limits – which is the reserve

Seismology the scientific study of earthquakes

Sequestration (in geology) usually used of carbon. The permanent removal of surface carbon into the subsurface geological environment

Shale a fine-grained, fissile sedimentary rock

Source rock a rock that is capable of producing hydrocarbons

Telemetry the process of recording and transmitting the readings of an instrument

Thermogenic formed by heat

Unconventional hydrocarbons oil and gas extracted from low-permeability rocks

Water table the upper limit of the saturated zone

CONVERSION TABLE

Measure	Conversion factor (multiply by)	Measure
Cubic feet of natural gas	0.0001767	Barrels of oil equivalent (boe)
Tonnes of oil equivalent (toe)	7.33	Barrels of oil equivalent (boe)
Barrels of oil (bbl)	0.1589873	Cubic metres
Barrels of oil (bbl)	34.97	UK gallons
Barrels of oil (bbl)	0.136	Tonnes of oil equivalent (toe)
UK gallons	4.545	litres
Pounds (lb)	0.45359237	kilograms
Miles	1.609344	kilometres
Feet (ft)	0.3048	Meters
Square miles	2.589988	Square kilometres
British Thermal Units (BTU)	1055.05585262	Joules (J)
Tonnes of oil equivalent (toe)	41.868	Gigajoules (GJ)
Cubic feet of natural gas	1025	British Thermal Units (BTU)
kilowatt hours (kWh)	3.6	Megajoules (MJ)
Tonnes of coal equivalent (tce)	29.39	Gigajoules (GJ)

INDEX

Note: Page numbers followed by *f* indicate figures.

Printed in the United States
By Bookmasters